Series in Mathematics Modules

MODULE XI

GRAPHING FUNCTIONS

GARY FITTS

University of California, Berkeley

Cummings Publishing Company

Menlo Park, California • Reading, Massachusetts
London • Amsterdam • Don Mills, Ontario • Sydney

To Celeste, Jessie, and Jake

The author wishes to thank Celeste Burrows and Leon Ablon for their criticisms and suggestions; George Wever, who contributed to the original version of this module; Alan Dankner and Jim Rosenberg, for their parts in the project; Martha Maxwell, for her constant support; and the staff of Cummings Publishing Company, who have been astonishingly tolerant and helpful.

Printed in the United States of America.
Published simultaneously in Canada.
Library of Congress Catalog Card Number 76-62884

ISBN: 0-8465-0265-8
ABCDEFGHIJK-AL-7987

Cummings Publishing Company, Inc.
2727 Sand Hill Road
Menlo Park, California 94025

PREREQUISITES

Module III *Linear Equations and Lines*
Module V *Quadratic Equations and Curves*
Module X *Functions and Word Problems*

TO THE STUDENT

To learn graphing, you've got to draw graphs—hundreds of them. This module tries to make it easy for you, but you have to remember to *read actively*. When you come to an example in the text, read the question and *cover up the solution*. Try to work it out for yourself first. There is a bit of artistry in graphing, and only practice will make it easy.

Your work will be easier still if you use a calculator. There is a lot of computation involved in graphing, and while you can certainly do it all by hand, even the cheapest calculator will speed up things considerably. If you are going to buy a calculator, look for one that has at least memory, square root, and reciprocal (1/x) keys. (Some other nice features are x^2, memory +, memory −, and additional memories. The nicest feature of all is programmability, but that's expensive. This module doesn't use sin, cos, or tan, but you might want them anyway.) Count on spending the first lesson in the book practicing with the calculator. Once you are proficient, arithmetic will be less apt to get in the way of the ideas you are trying to learn, and you will be more apt to take the time to read actively.

CONTENTS

Lesson 1

GRAPHS

REVIEW

When we graph an equation such as $y = x^2 - 5x + 4$, what we are really doing is drawing a picture of the solutions of the equation. So first we make a table of solutions and then we plot these solutions as points on a graph.

$$y = x^2 - 5x + 4$$

x	$x^2 - 5x + 4$	y	plot (x,y)
-1	$(-1)^2 - 5(-1) + 4$	10	(-1,10)
0	$(0)^2 - 5(0) + 4$	4	(0,4)
1	$(1)^2 - 5(1) + 4$	0	(1,0)
2	$(2)^2 - 5(2) + 4$	-2	(2,-2)
3	$(3)^2 - 5(3) + 4$	-2	(3,-2)
4	$(4)^2 - 5(4) + 4$	0	(4,0)
5	$(5)^2 - 5(5) + 4$	4	(5,4)
6	$(6)^2 - 5(6) + 4$	10	(6,10)

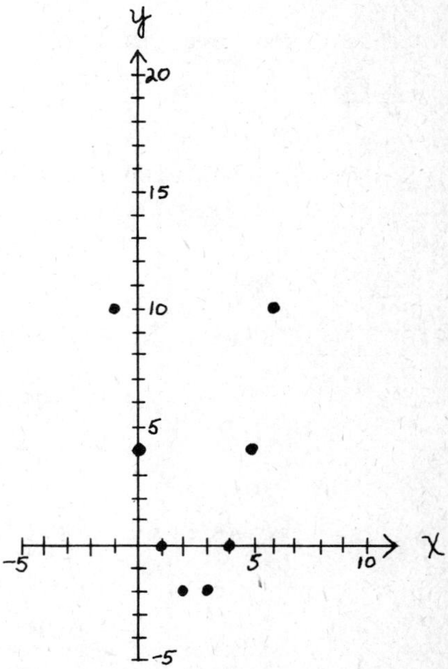

Then we fill in other solutions by drawing a curve through these points, getting a curve called a parabola.

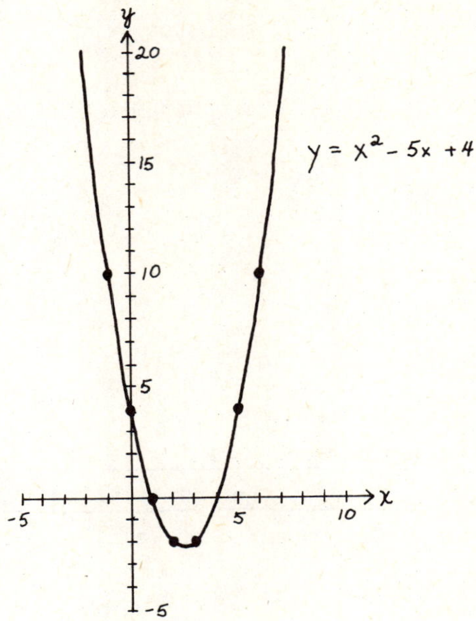

(This was covered in lesson 4 of Module V, Quadratic Equations and Curves.)

So far this has been review. Now let's look at the same example from a slightly different point of view.

GRAPHING RULES

EXAMPLE 1 Let's graph the following rule (or "function").

rule F: Take the input, square it, subtract 5 times the input, and add 4. The result is the output.

or

rule F: $F(x) = x^2 - 5x + 4$

Notice the similarity between this rule and the equation in the review.

$$F(x) = x^2 - 5x + 4$$
$$y = x^2 - 5x + 4$$

It should not be surprising that the graph of the rule turns out to be exactly the same as the graph of the equation, except that we use slightly different words to describe them. First we make a table of inputs and outputs:

$F(x) = x^2 - 5x + 4$

input	computation	output	plot (input,output)
-1	$(-1)^2 - 5(-1) + 4$	10	(-1,10)
0	$(0)^2 - 5(0) + 4$	4	(0,4)
1	$(1)^2 - 5(1) + 4$	0	(1,0)
2	$(2)^2 - 5(2) + 4$	-2	(2,-2)
3	$(3)^2 - 5(3) + 4$	-2	(3,-2)
4	$(4)^2 - 5(4) + 4$	0	(4,0)
5	$(5)^2 - 5(5) + 4$	4	(5,4)
6	$(6)^2 - 5(6) + 4$	10	(6,10)

Next we plot the (input,output) pairs of numbers exactly the same way that we plotted the (x,y) pairs. But notice that we label the vertical axis with the symbol F(x) instead of with y.

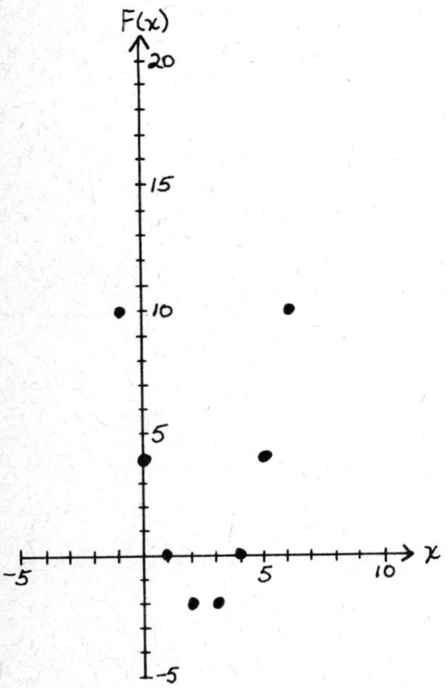

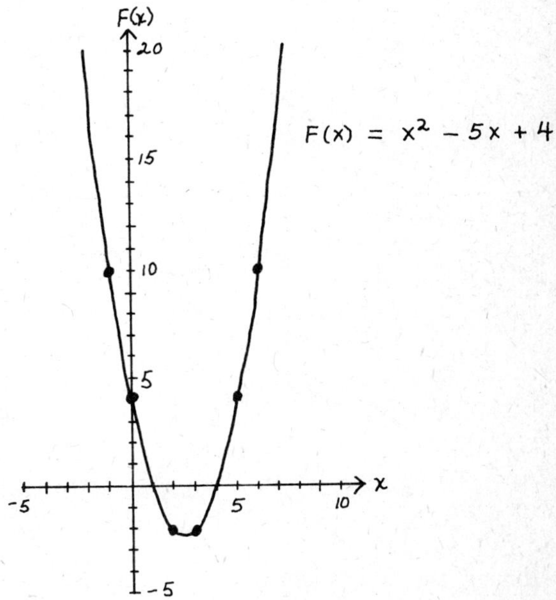

Then we draw a curve through these points.

Except for the label, this graph looks just like the previous one. It gives us lots of (input,output) pairs. For example, (6,10) is a point on the graph. (Actually it is one of the points that we plotted.) This tells us that

For $y = x^2 - 5x + 4$, when x is 6, y is 10.
For $F(x) = x^2 - 5x + 4$, when the input is 6, the output is 10.

By looking at the graph, can you tell the output of rule F given input -2?

Answer: By using a ruler or any straight edge we can see that when the input is -2, the output is 18.

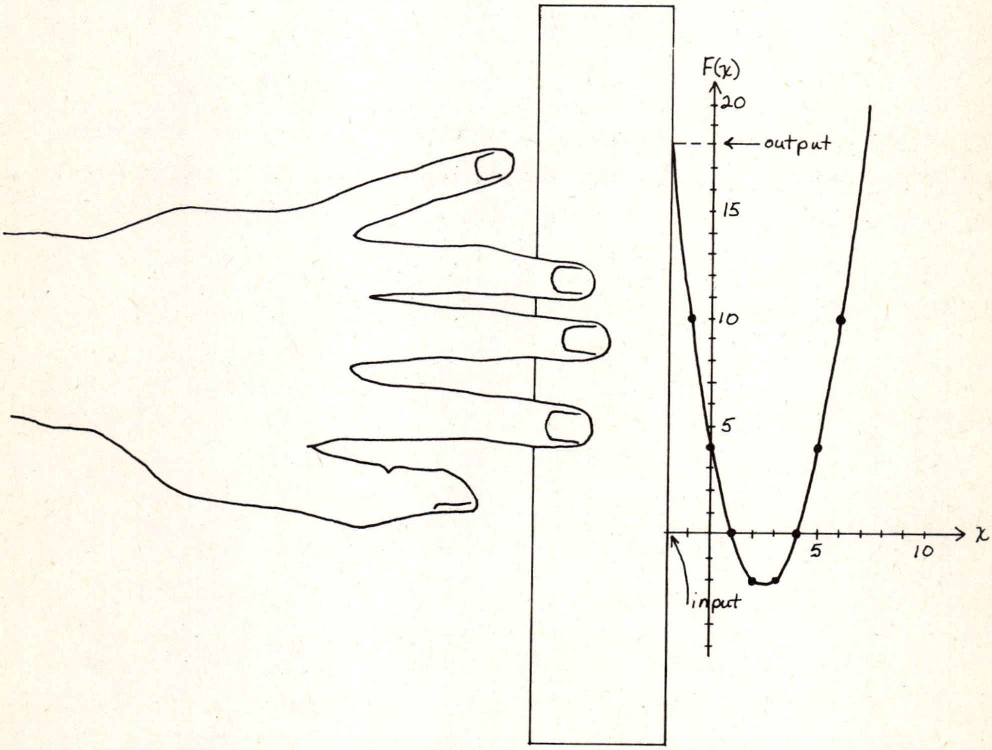

Using the graph again, find the output of rule F given input 7. Your answer should be 18. Actually, since the accuracy with which we can

read graphs is limited, any answer close to 18 (like 17.3 or 18.7) is OK. Of course, we can get the exact answer by doing the computation:

$$F(-2) = (-2)^2 - 5(-2) + 4 = 18.$$

By now you have seen lots of graphs that are straight lines or parabolas. But as you might suspect, there are many other kinds of curves.

<u>EXAMPLE 2</u> Graph $F(x) = x^3 - 3x$.

input	computation	output	plot
0	$0^3 - 3 \cdot 0$	0	(0,0)
1	$1^3 - 3 \cdot 1$	-2	(1,-2)
2	$2^3 - 3 \cdot 2$	2	(2,2)
3	$3^3 - 3 \cdot 3$	18	(3,18)

Notice that we chose some easy inputs to begin with. We may not have enough, but let's plot these (input,output) pairs and see if we can tell what the curve is going to look like.

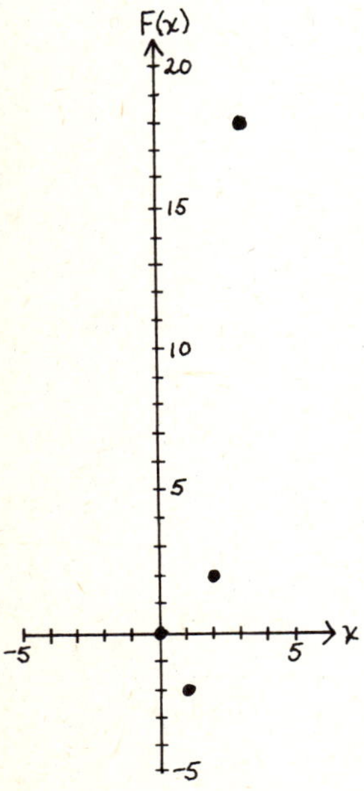

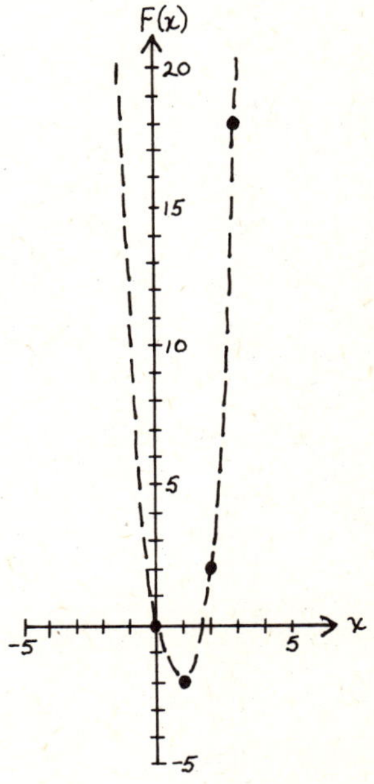

wrong

Our first guess might be a curve that looks a lot like a parabola; but as we shall see, this is wrong.

Let's calculate some more (input,output) pairs for this rule and plot them on the graph above. If these new points are on the curve, then our curve is probably right. But if they are off the curve, then it must be wrong.

input	computation	output	plot
-1	$(-1)^3 - 3(-1)$	2	(-1,2)
-2	$(-2)^3 - 3(-2)$	-2	(-2,-2)
-3	$(-3)^3 - 3(-3)$	-18	(-3,-18)

Plot these points on the last graph. They don't fit the curve at all! In fact (-3,-18) doesn't even fit on the picture. So let's draw a new set of axes that will let us plot all the points.

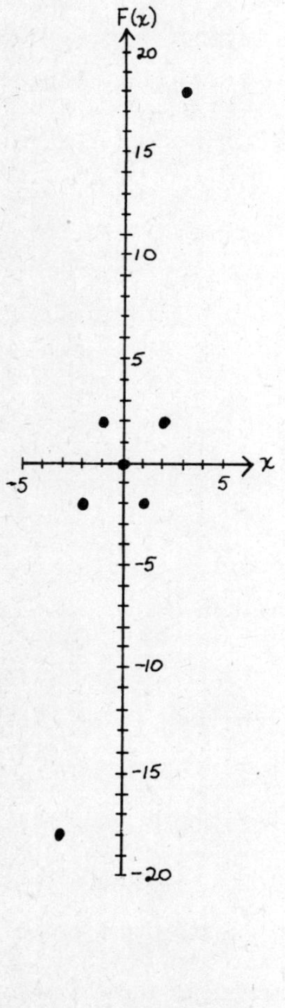

Can you tell what the curve will look like from these seven points? Take your pencil and try connecting the points from left to right to make a smooth curve. (Don't worry about getting it wrong. Try it!)

If you have any doubts about the curve you just drew, the only way to check it is to plot some more points! Unfortunately, we have run out of easy inputs. We'll have to use fractional numbers. Let's use inputs halfway between some of those we have already plotted: -1.5, -0.5, 0.5, 1.5. There is a bit of computation to do, but you should work through the table until you are sure you can do it.

$F(x) = x^3 - 3x$

input	computation	output	plot
-1.5	$(-1.5)^3 - 3(-1.5)$	1.125	(-1.5,1.1)
-0.5	$(-0.5)^3 - 3(-0.5)$	1.375	(-0.5,1.4)
0.5	$(0.5)^3 - 3(0.5)$	-1.375	(0.5,-1.4)
1.5	$(1.5)^3 - 3(1.5)$	-1.125	(1.5,-1.1)

(Notice that we rounded off the output 1.125 to 1.1 and 1.375 to 1.4 since on this graph our eyes can't tell the difference.)

Plot these points on the last graph. Do the new points lie on the curve that you drew? If they do, then you drew it very accurately. Otherwise, erase the old curve and draw a new one through all the points. Then compare your sketch with the curve on the right side of the next page:

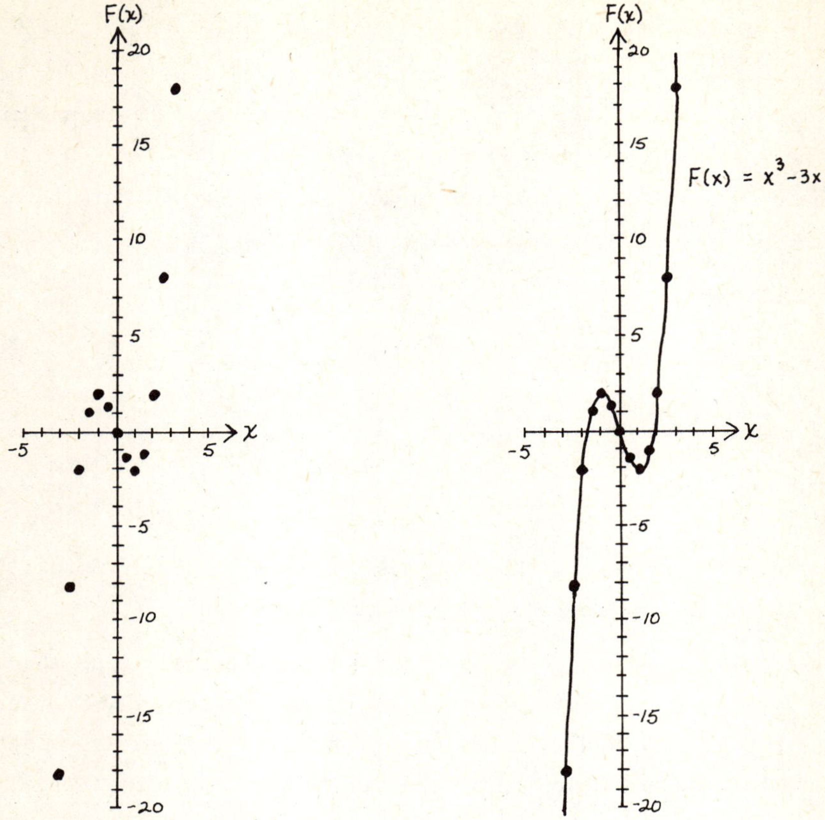

<u>Note on accuracy of hand-drawn graphs</u>: In this module we have purposefully drawn the graphs "realistically" as a guide to how your own graphs should look. We could have used drafting equipment and produced microscopically accurate illustrations, but we want to emphasize that this is <u>not important</u> for the graph sketching that you will do as a mathematics student. <u>The most important thing is the shape of the curve</u>. As you work through this module you will learn more details about what to watch out for when sketching graphs by hand. For now, use our sketches as a guide, and concentrate on getting the right shape.

S-shaped curves like the last one are called <u>cubics</u> when they come from formulas like $F(x) = x^3 - 3x$, where the highest power of x is x^3. (Remember that x^3 is read <u>x cubed</u>.)

The moral to this last example is: Whenever you're drawing a new graph that you haven't seen before, check it! In calculus, you will learn some shortcuts for checking graphs. But often the easiest way to check them is to plot a few new points and see if they fit on the curve you've already drawn.

EXAMPLE 3 Graph $F(x) = 3x - x^3$.

Remember, the important thing is the shape of the curve. As long as we get the right shape, we needn't worry too much about precision. The steps to follow are listed below. Follow these steps to sketch the graph, and then compare your sketch with the answer that follows.

- Choose some easy inputs and make a table of (input,ouput) pairs.
- Look at your table and see what the biggest numbers are, so you will know how long to make the x-axis and the F(x)-axis (the y-axis).
- Plot the (input,output) pairs.
- Connect these points from left to right with a smooth curve.
- Plot a few new points as a check, and change your graph if necessary.

Answer:

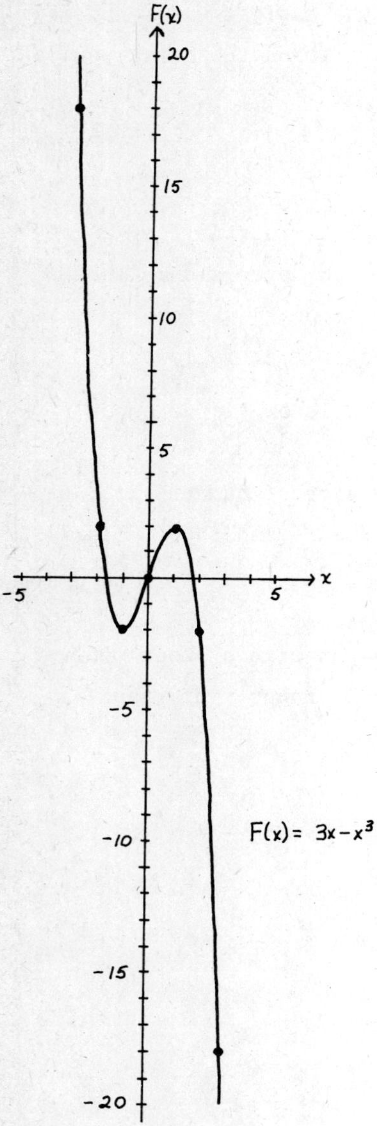

Do you notice any similarity between this graph and the one in example 2?

EXAMPLE 4 Graph F(x) = 2x - 3, and G(x) = -2.

F(x) = 2x - 3

input	output
0	-3
2	1

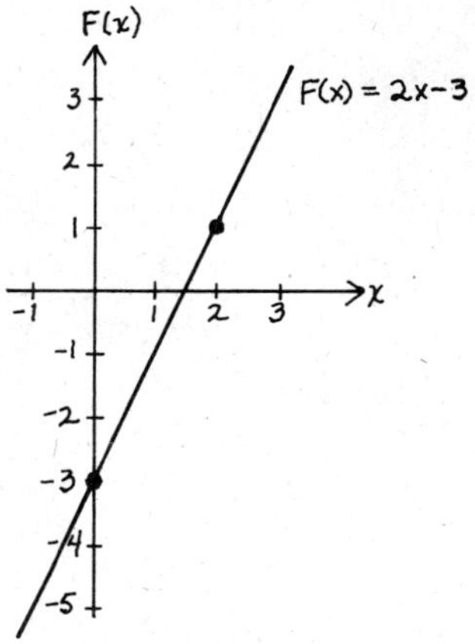

G(x) = -2

input	output
-3	-2
2	-2

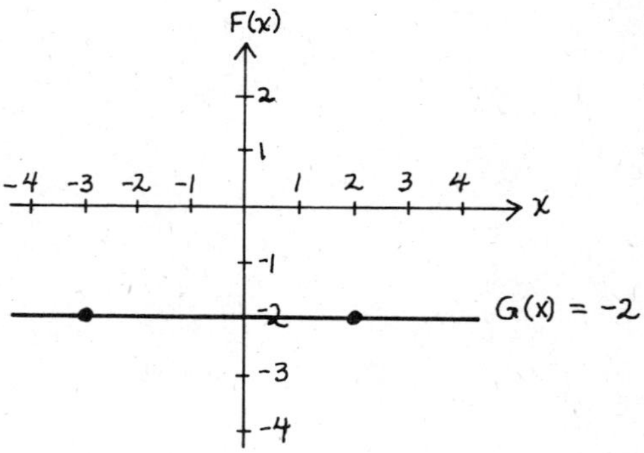

Note: You probably know another way to graph straight lines. Module III of this series, Linear Equations and Lines, explained how to graph lines by the intercept-slope method. Use this technique whenever it is convenient. Remember that point-plotting is a general technique that you can always fall back on; it works for any rule. But the intercept-slope method works only for straight lines.

We used point-plotting in the review and in example 1 to graph a parabola, although you probably know a shortcut here also. As long as the parabola crosses the x-axis, we can plot the crossing points. Then we can use an input halfway between these to plot the top (or bottom) point, and we can draw the parabola using only these three points. This was covered in Module V, Quadratic Equations and Curves. Use this technique whenever it is convenient. But remember that it applies only to parabolas—the graphs of formulas where the highest power of x is x^2—and not to other kinds of graphs.

EXERCISES (Answers start on page 14.)

Graph the following rules.

1. $F(x) = 2x + 1$
2. $G(x) = \frac{x}{2} + 1$
3. $H(x) = -\frac{x}{2} + 1$
4. $F(x) = 3$
5. $G(x) = -3$
6. $F(x) = x^2$
7. $G(x) = x^2 + 4$
8. $H(x) = x^2 - 4$
9. $F(x) = x^2 - 4x$
10. $G(x) = 4x - x^2$
11. $F(x) = \frac{x^3}{6}$
12. $G(x) = \frac{x^3}{6} + 2$
13. $H(x) = \frac{x^3}{6} + x$
14. $F(x) = \frac{x^3}{6} - x$
15. $G(x) = x - \frac{x^3}{6}$

ANSWERS TO EXERCISES

(1)

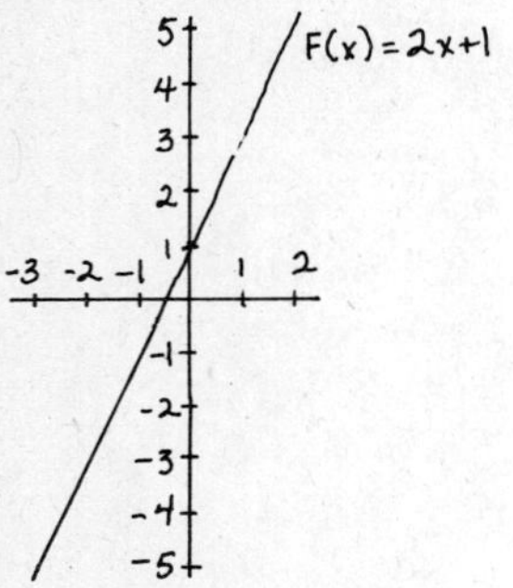

(2)

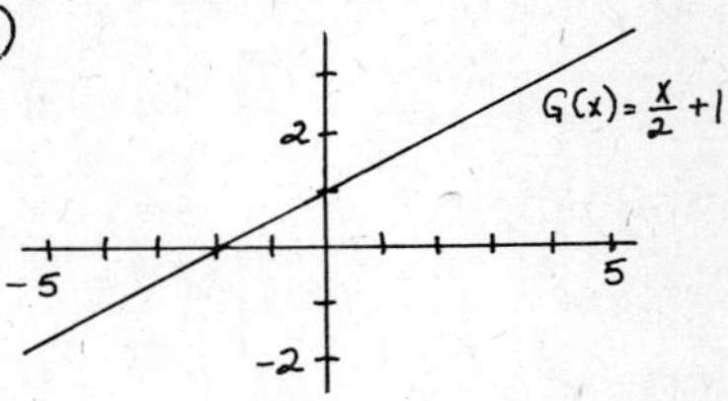

(3)

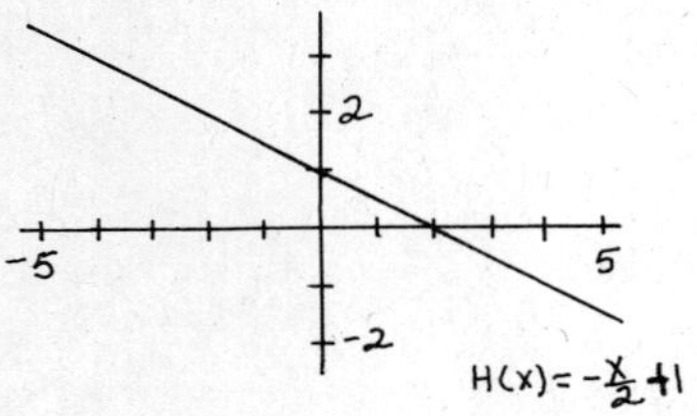

(4)

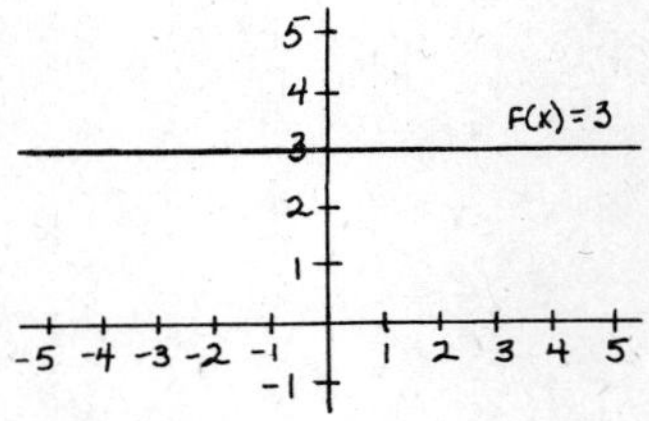

(5)

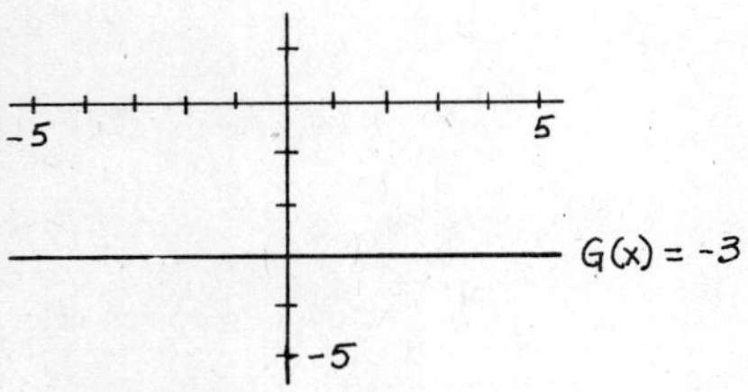

(6)

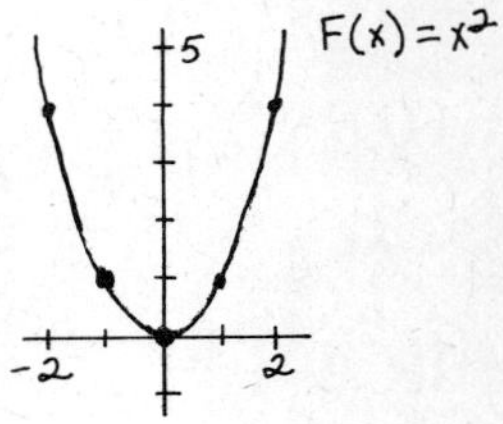

(7)

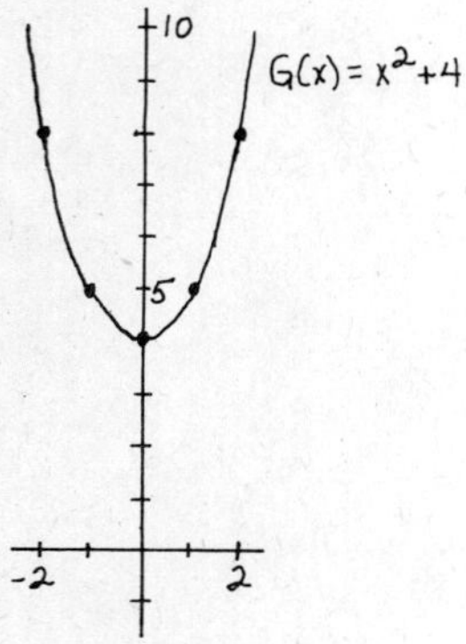

(8)

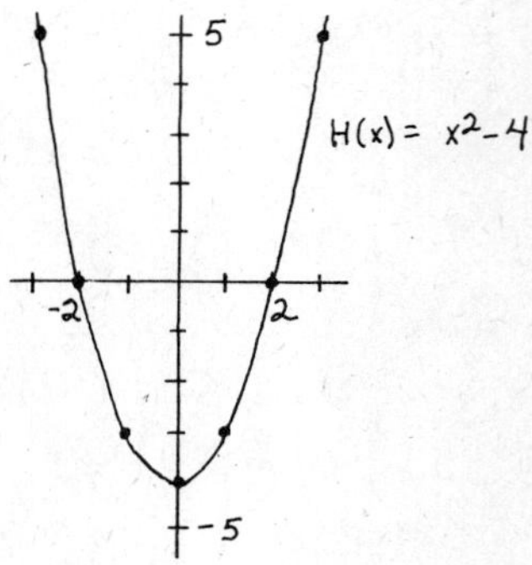

(9.)

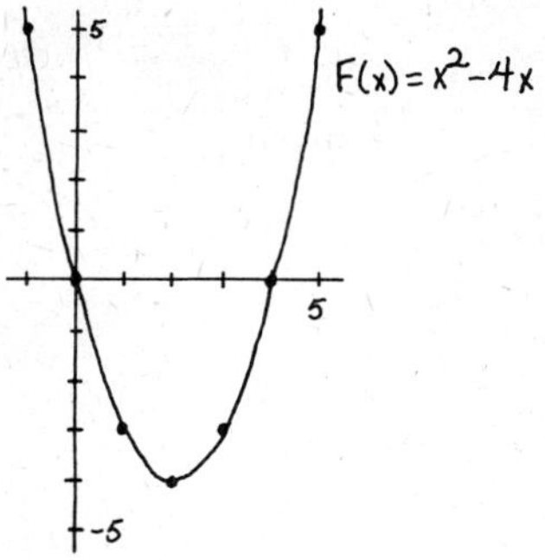

(10)

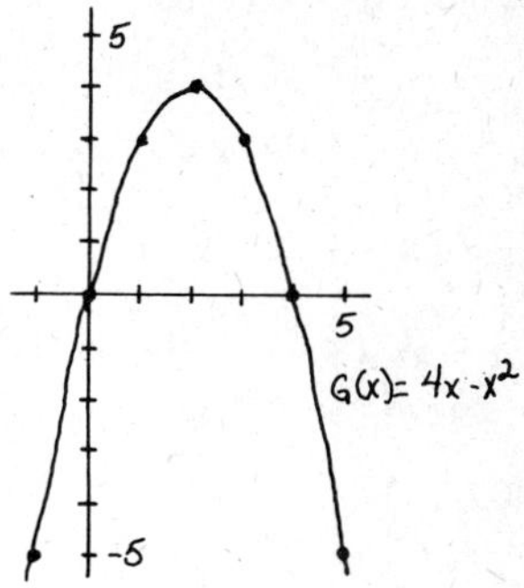

(11.)

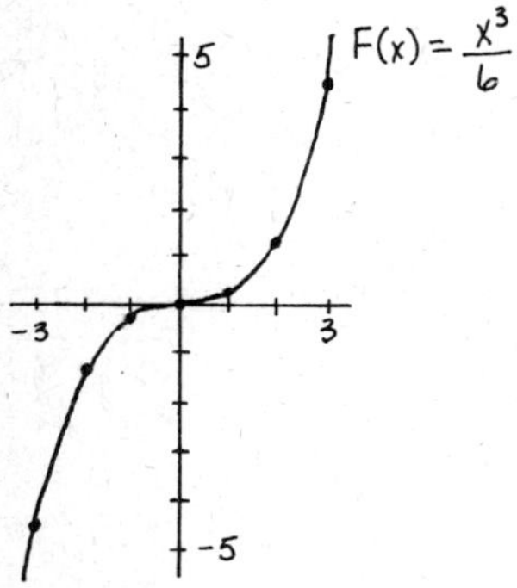

(12)

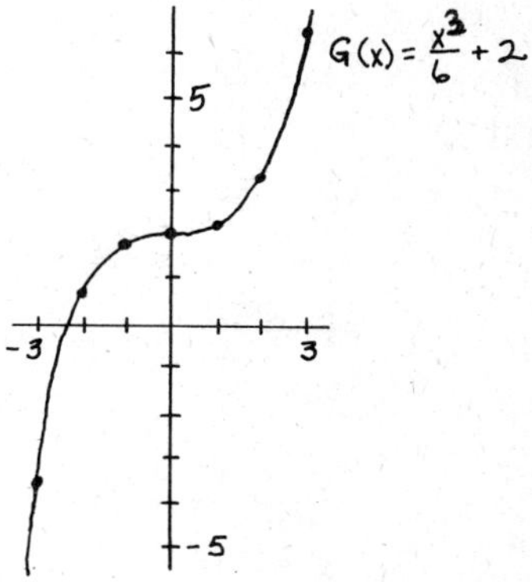

(13)

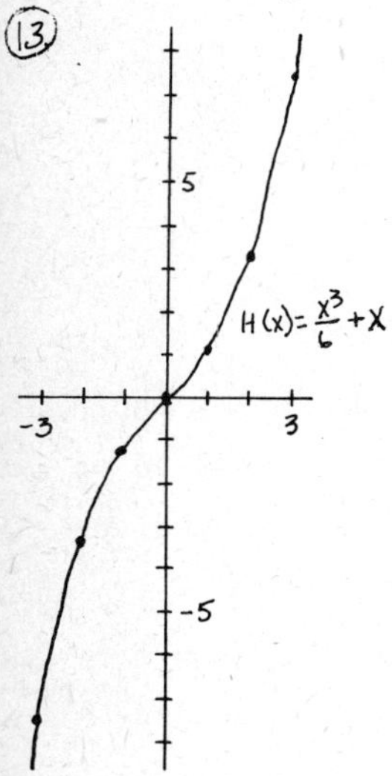

(14)

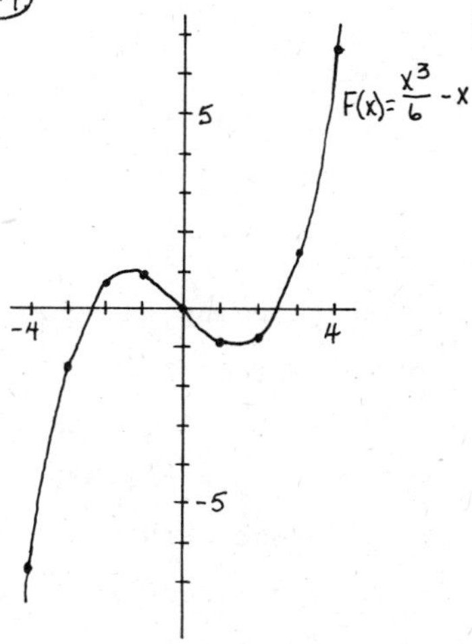

(15)

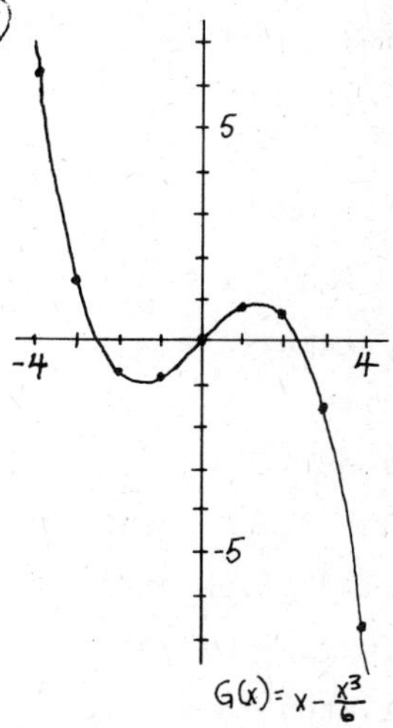

ADDITIONAL EXERCISES

Graph the following rules.

1. $F(x) = 2x - 1$
2. $G(x) = \frac{x}{2} + 2$
3. $H(x) = -\frac{x}{2} + 2$
4. $F(x) = 4$
5. $G(x) = -4$
6. $F(x) = (x - 2)^2$
7. $G(x) = (x - 2)^2 + 4$
8. $H(x) = (x - 2)^2 - 4$
9. $F(x) = (x - 2)^2 - 4(x - 2)$
10. $G(x) = 4(x - 2) - (x - 2)^2$
11. $F(x) = \frac{(x - 2)^3}{6}$
12. $G(x) = \frac{(x - 2)^3}{6} + 2$
13. $H(x) = \frac{(x - 2)^3}{6} + (x - 2)$
14. $F(x) = \frac{(x - 2)^3}{6} - (x - 2)$
15. $G(x) = (x - 2) - \frac{(x - 2)^3}{6}$

SUMMARY OF LESSON 1

Graphing a rule is just like graphing an equation. We follow the same steps, except that we use the words "input" and "output" instead of "x" and "y." We plot the points exactly the same way, and draw a curve through these points in the same way [pages 1-5].

The graph of a quadratic rule (where the highest power of x is x^2) is always a parabola. The graph of a linear rule (where the rule can be written in the form $F(x) = mx + b$) is always a straight line [page 11]. But of course, there are many other kinds of curves. This lesson introduces cubic rules (where the highest power of x is x^3).

Their graphs are s-shaped curves such as 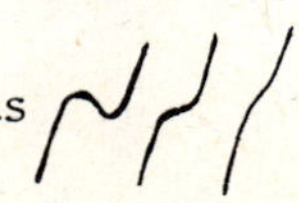 or 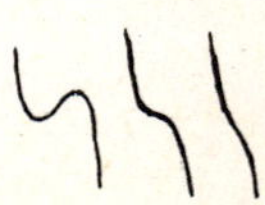 [pages 5-10].

The steps to follow in sketching these graphs are summarized on page 9. Some shortcuts are mentioned on pages 11-12.

If you have any trouble with any of the exercises, and the answers don't provide enough help, use the following as a guide for review.

exercises	review
1-5	example 4, page 11; page 12; Module III
6-10	review and example 1, pages 1-5; pages 11-12; Module V
11-15	examples 2 and 3, pages 5-10

Lesson 2

GRAPH CHECKING AND USING DIFFERENT SCALES

In lesson 1, we showed how to check a graph by plotting a few new points and seeing whether or not they lie on the curve. We should check graphs carefully where they bend most sharply. In particular, we should check them near the "peaks" $\cap$ and "valleys" $\cup$ of the curve.

EXAMPLE 1 Graph $F(x) = x(2x - 3)^2$.

First choose some easy inputs and make a table. This time we won't show all the computations. But you should check at least a few; check all of them if you have a calculator.

$F(x) = x(2x - 3)^2$

input	output	plot
-3	-243	
-2	-98	
-1	-25	(-1,-25)
0	0	(0,0)
1	1	(1,1)
2	2	(2,2)
3	27	(3,27)
4	100	
5	245	

Actually we see that some of these inputs were not so "easy." They are all easy to compute, but outputs like -243 are harder to graph. We will come back to these shortly, but for now let's just plot the easiest ones. (It would be good practice to sketch the curve, and check it, before you look at our sketch. In particular, check your graph where it bends most sharply.)

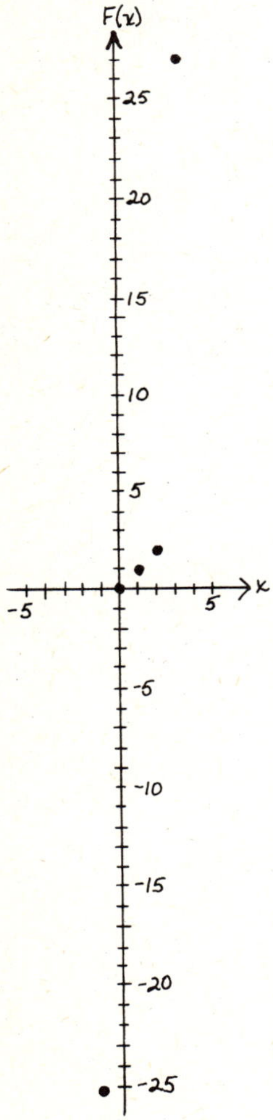

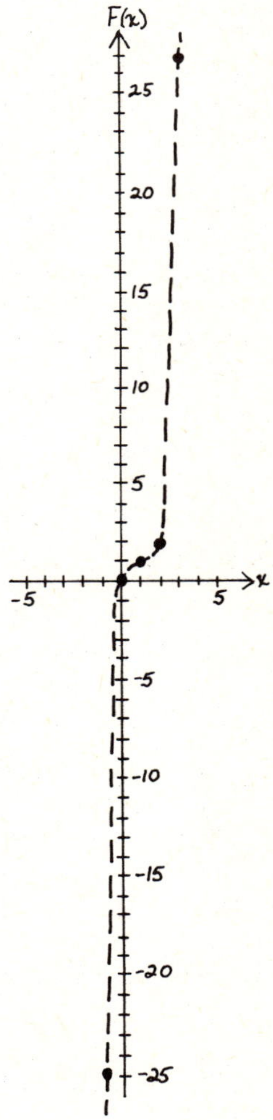

wrong

The graph on the left shows the points, and the one on the right shows a curve that we might have drawn at first. But let's see what happens when we check this graph where it bends most sharply.

$F(x) = x(2x - 3)^2$

input	output	plot
0.5	2	(0.5,2)
1.5	0	(1.5,0)

These new points don't fit the curve!

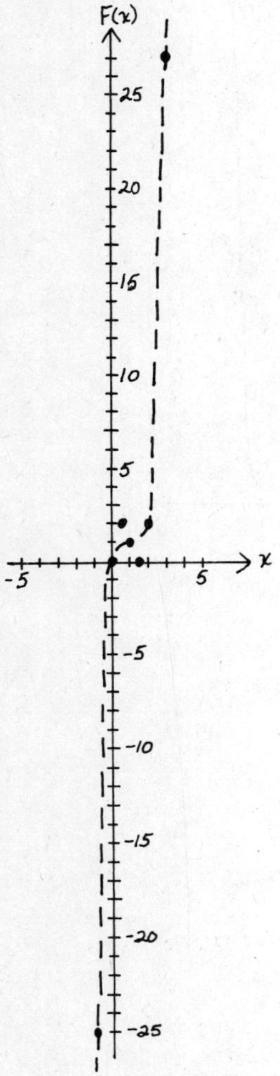

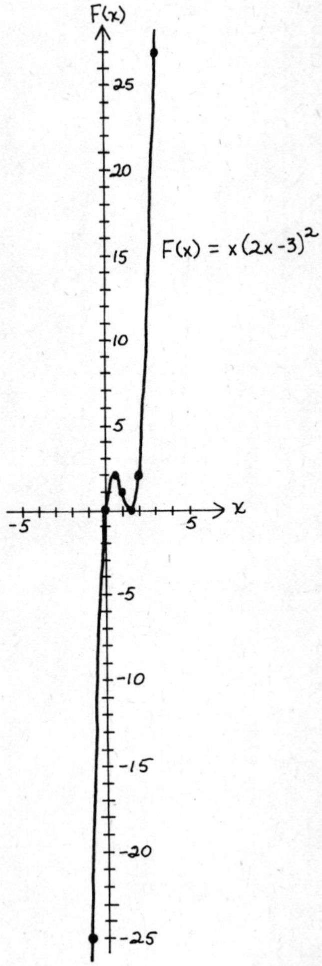

The graph on the left shows all the (input,output) pairs that we have figured so far. Connecting them from left to right with a smooth curve gives the graph on the right.

Notice that the graph is cubic, since multiplying $x(2x - 3)^2$ gives $x(4x^2 - 12x + 9)$, which is $4x^3 - 12x^2 + 9x$.

But how can we be sure that we finally have the right shape for this graph?

We could plot some more (input,output) pairs around the peaks and valleys of this curve, using inputs between the ones we already have. But that would mean using inputs like 0.2 and 1.8, and these would be hard to find accurately on our small x-axis. So we can expand the x-axis as follows:

USING DIFFERENT SCALES

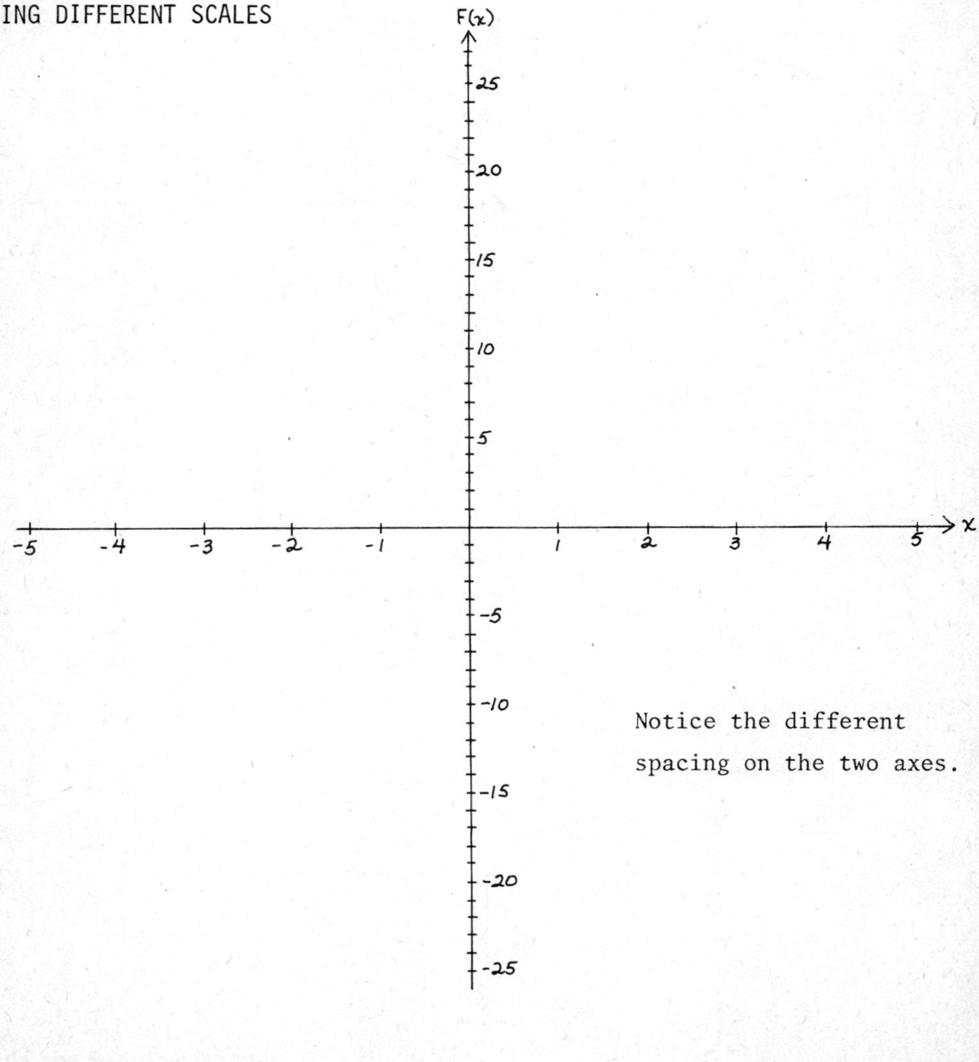

Notice the different spacing on the two axes.

Now we can re-plot our (input,output) pairs and draw a curve through them much more easily.

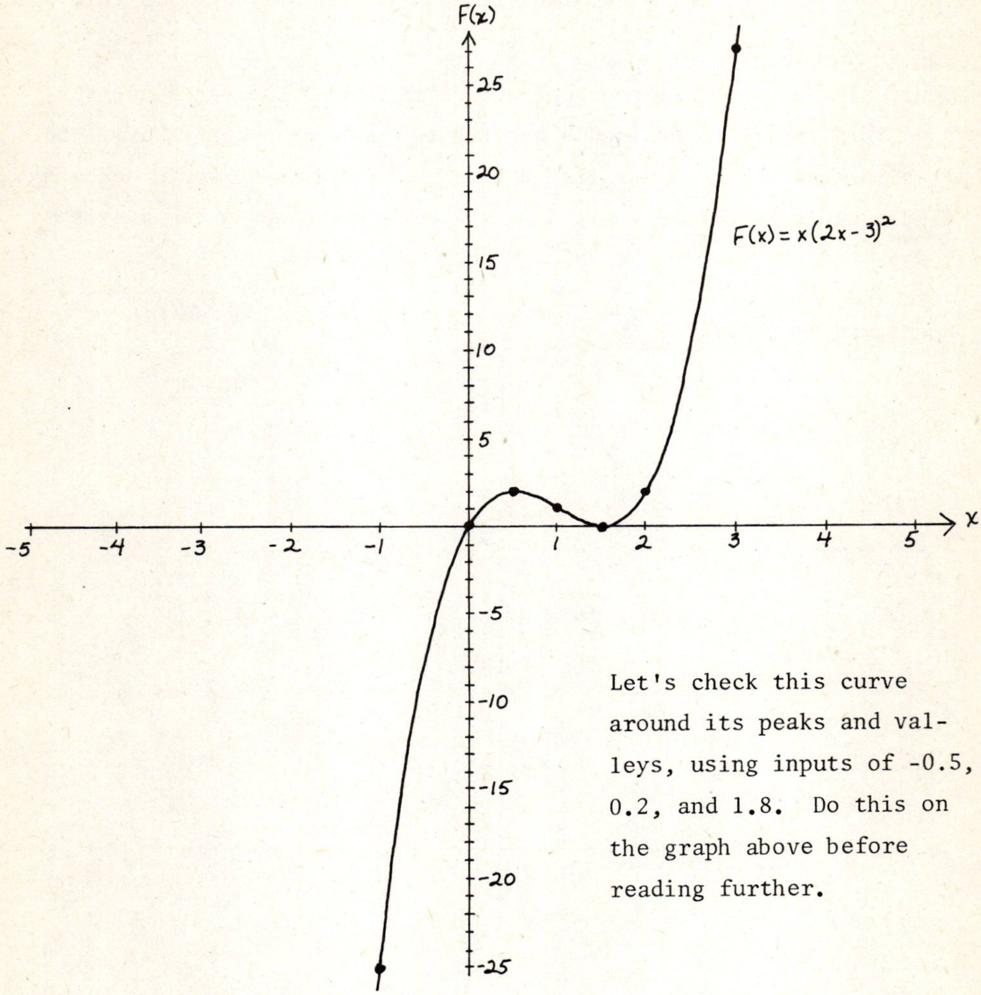

Let's check this curve around its peaks and valleys, using inputs of -0.5, 0.2, and 1.8. Do this on the graph above before reading further.

$F(x) = x(2x - 3)^2$

input	output	plot
-0.5	-8	(-0.5,-8)
0.2	1.352	(0.2,1.4)
1.8	0.648	(1.8,0.6)

These new (input,output) pairs are plotted below, and they do lie on the curve. Notice that we rounded off the output 1.352 to 1.4 and 0.648 to 0.6 since on this graph our eyes can't tell the difference.

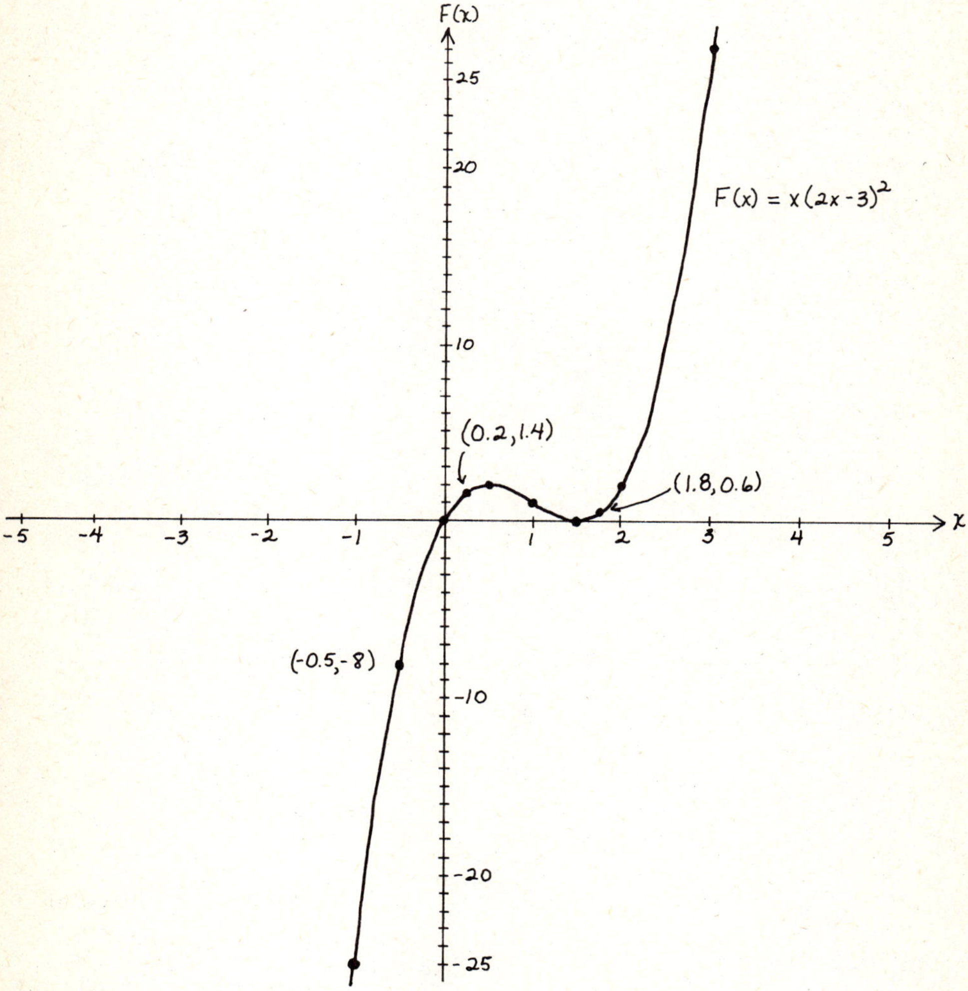

Finally, is there any way we can plot those "far-out" pairs that we calculated at the beginning of this example? These were (-3,-243), (-2,-98), (4,100), and (5,245). We can do it by re-numbering the F(x)-axis as follows:

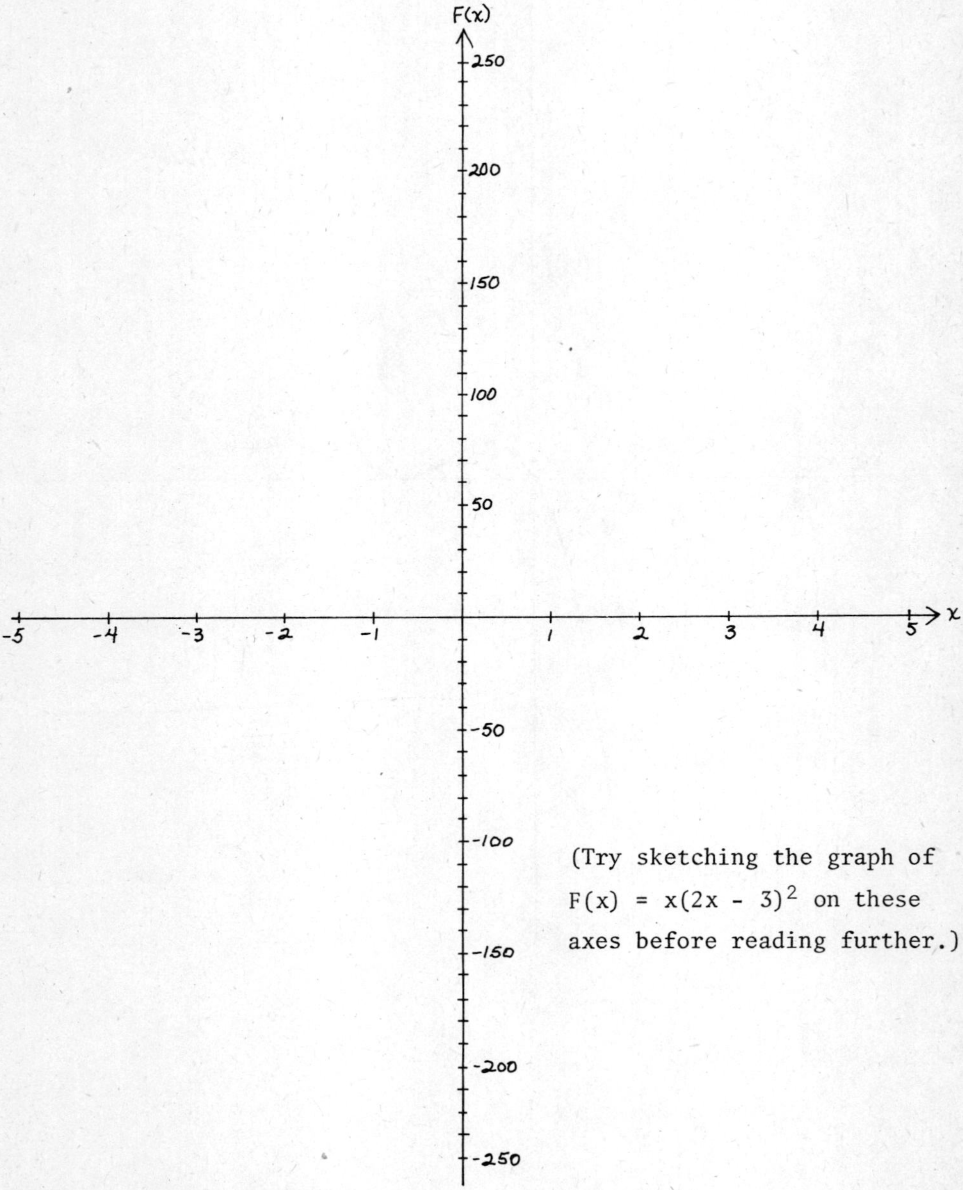

(Try sketching the graph of $F(x) = x(2x - 3)^2$ on these axes before reading further.)

Now when we plot all our (input,output) pairs, it's very hard to plot the smaller numbers. The curve hardly appears to rise above the x-axis at all between 0 and 2, even though we know it does. But at least the "far-out" points fit nicely on the curve.

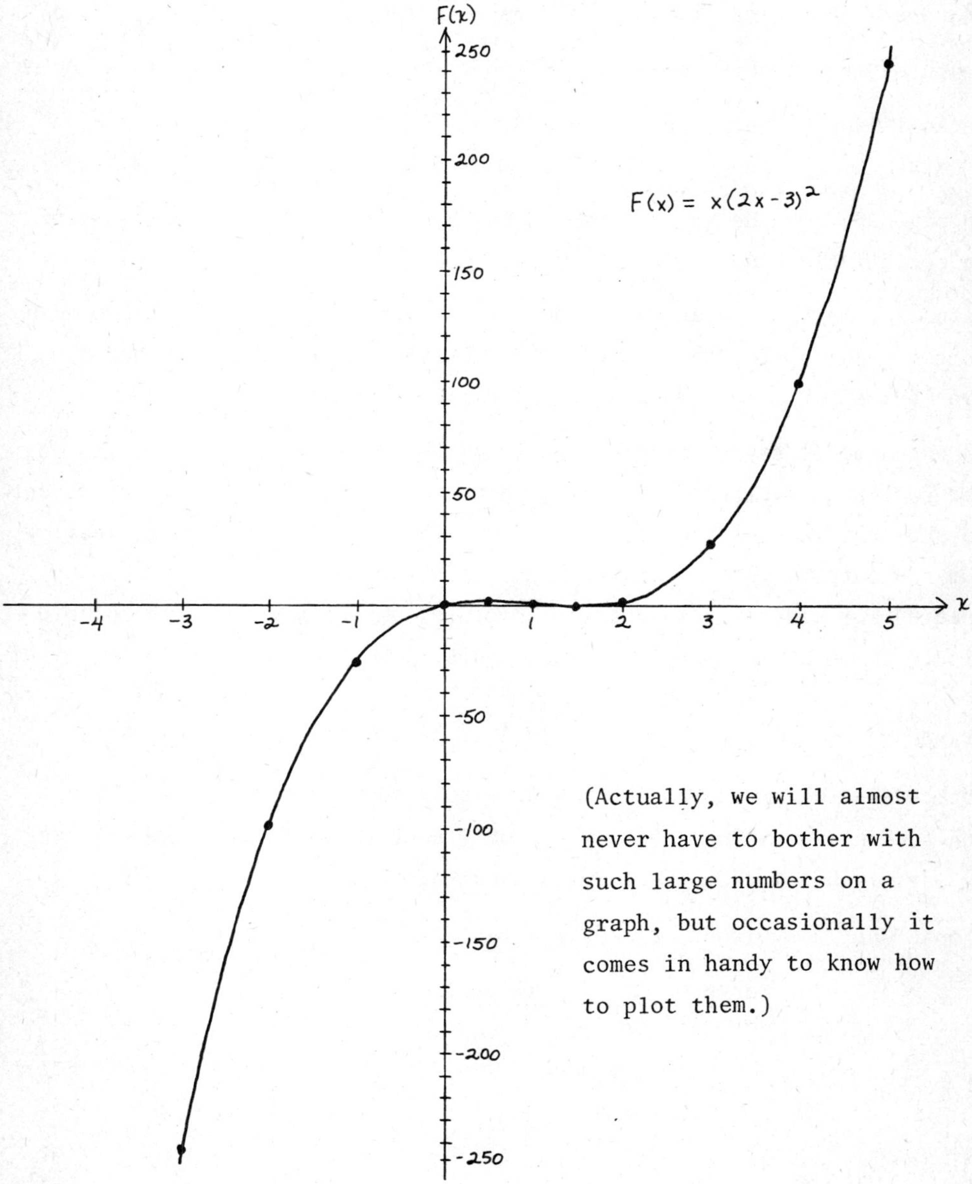

(Actually, we will almost never have to bother with such large numbers on a graph, but occasionally it comes in handy to know how to plot them.)

Note: In this example we used different spacing on the x-axis and F(x)-axis in order to make it easier to plot points. Remember this technique. Some graphs would be impossible to draw without it.

We used this technique for two different purposes in the last example. On page 22, we used it to get a more detailed look at the part of the graph where the curve makes its sharpest bends . We had to use decimal inputs there, and it helped to use a bigger spacing on the x-axis.

On page 25, we made the spacing on the F(x)-axis much smaller, so that we could include some "far-out" outputs.

Actually, mathematicians are usually much more interested in the bending parts of the curve, the "peaks"  and the "valleys" , than they are in the parts that look almost straight.

So when you're trying to decide how to number the x- and F(x)-axes, do the following: First decide what inputs and outputs to include in order to show the interesting parts of the curve. This usually means leaving out any outputs that are much larger than the others. Then choose spacing for each axis so that the picture comes out to be a reasonable size. The next example illustrates this.

EXAMPLE 2 Graph $F(t) = t(t + 1)(8t^2 - 10t - 9)$.

Notice that we are using the letter t to stand for the input. This won't change anything; we always plot (input,output) pairs the same way, no matter what letter the formula uses for the input.

Choose some easy inputs and make a table.

input	output	plot
-2	86	Leave this one out for now.
-1	0	(-1,0)
0	0	(0,0)
1	-22	(1,-22)
2	18	(2,18)
3	396	Leave this one out for now.

We can see that our t-axis should include inputs from -1 to 2, and our F(t)-axis should include outputs from -22 to 18.

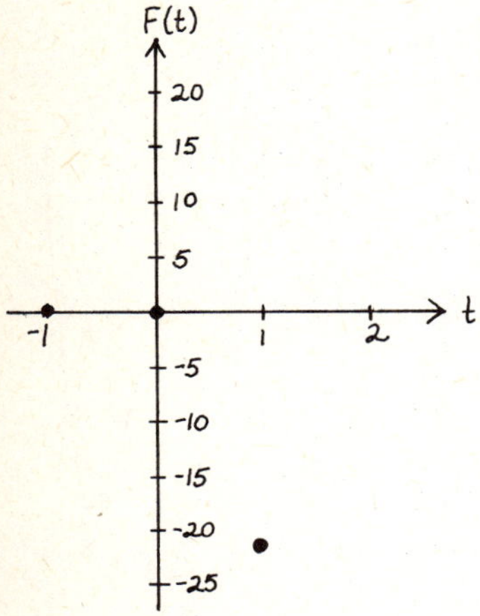

Notice that we chose the spacing on the axes to make the picture come out a reasonable size. Compare this with the graphs on pages 19-24, which are bigger than necessary.

Do we have enough information yet to draw the graph? If we think about the two points we left out, we can be fairly sure of the left and right sides of the curve.

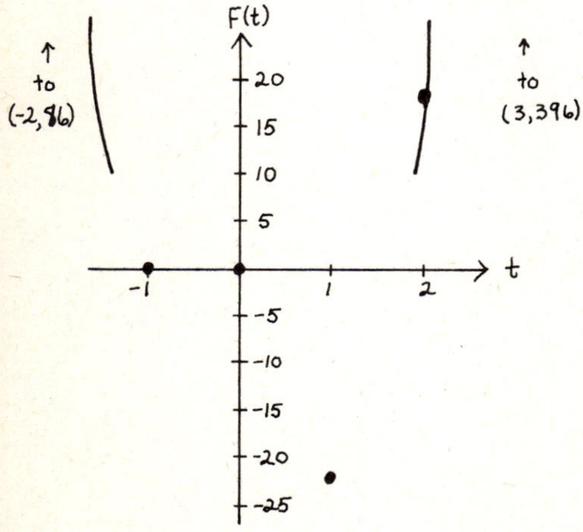

We know that the left side of the curve looks this way, because even though we didn't plot the point (-2,86), we can imagine where it is and draw the curve in that direction. We can do the same for (3,396) on the right side.

But there are still many possibilities for the most interesting parts of the curve, where it makes all its turns and bends. A few of these are shown below.

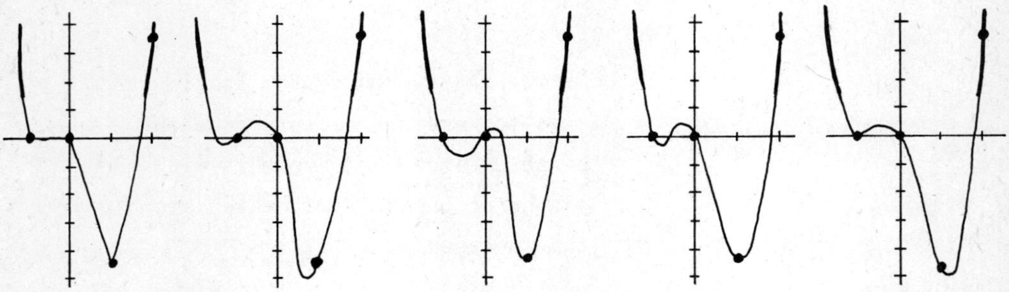

(a) (b) (c) (d) (e)

We can rule out possibility (a) because it has sharp corners—almost every rule has a smooth graph. But there are still many other possibilities around the peaks and valleys, so we should plot some more points.

$F(t) = t(t + 1)(8t^2 - 10t - 9)$

input	output	plot
-0.5	0.5	(-0.5,0.5)
0.5	-9	(0.5,-9)
1.5	-22.5	(1.5,-22.5)

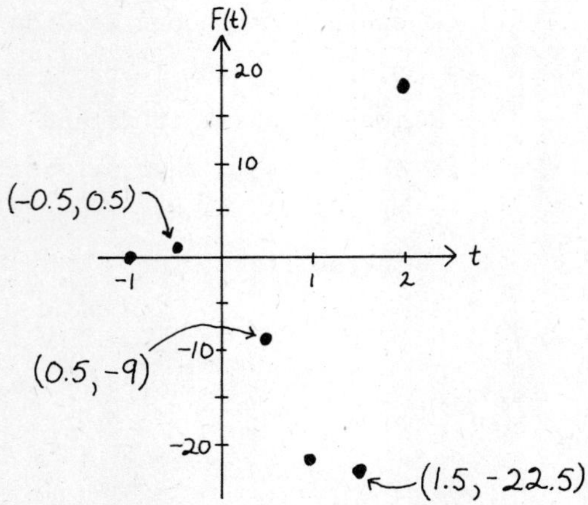

Now we can rule out possibility (c). But there are still three reasonable possibilities for what happens to the left of the F(t)-axis.

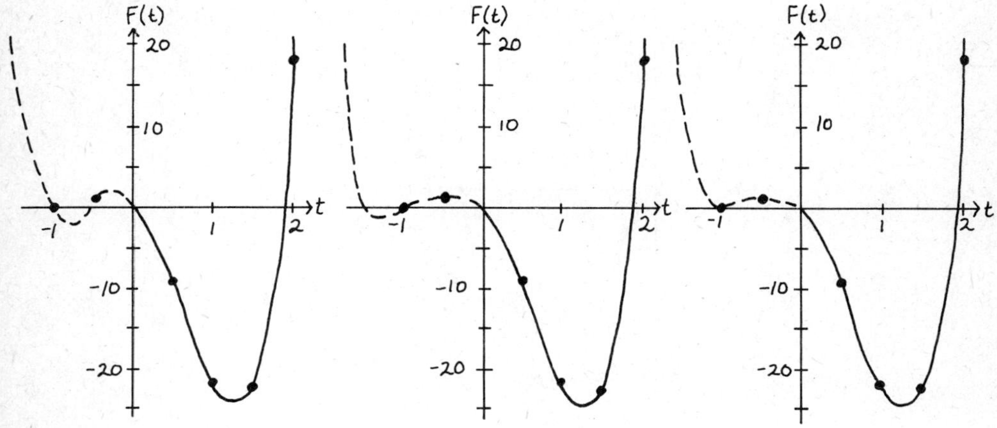

Let's try some inputs close to -1, say -1.2 and -0.8. The new (input, output) pairs are (-1.2,3.4848) and (-0.8,-0.6592), which we plot as (-1.2,3.5) and (-0.8,-0.7).

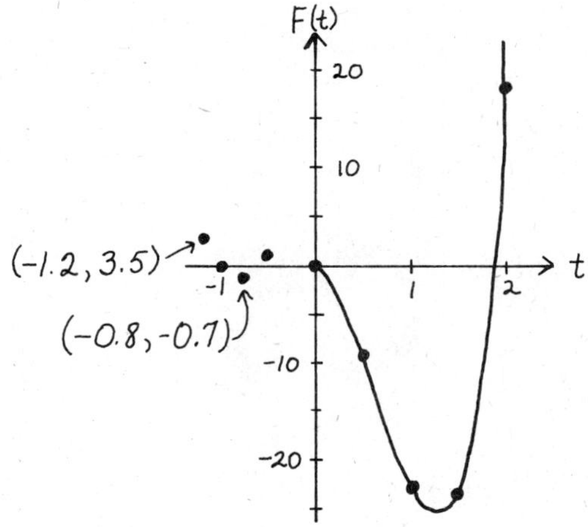

Finally, we can be fairly sure of the shape.

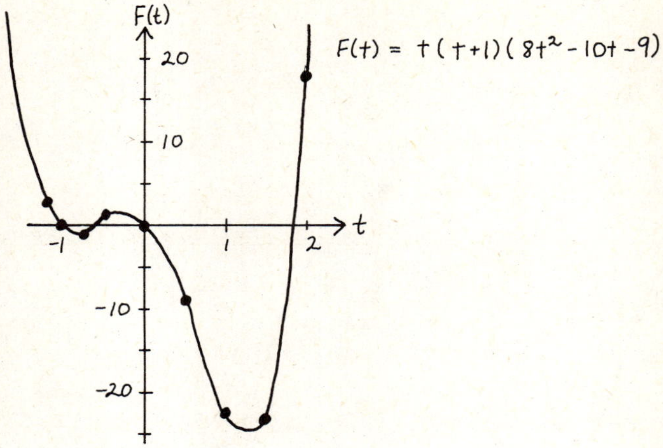

Note: At the beginning of this example we calculated the (input,output) pair (-2,86). Even though we didn't plot it, we "imagined" where it should be in order to decide which way to draw the left side of the curve. Without this, we might have made the mistake of drawing the curve as though it were a cubic:

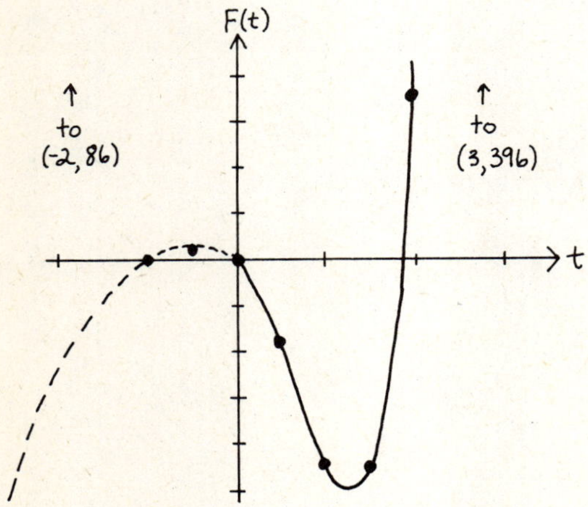

wrong

In this lesson, we have added two important steps to the list of steps in lesson 1 for sketching graphs.

1. Plot some new (input,output) pairs where the curve bends most sharply, and correct the curve if necessary.
2. Calculate some "far-out" (input,output) pairs, at least one on each side of the curve, and imagine where they are located to see if the curve is heading in the right direction.

Here is a new summary of the steps to follow in sketching a graph.

√ Choose some easy inputs and make a table of (input,output) pairs.

√ Decide which pairs you are going to plot. For now, disregard outputs that are much larger than the others.

√ Draw a line for the input-axis. Look at the (input,output) pairs that you are going to plot, and decide what range of inputs to include on this axis. Number this axis, using a spacing that makes it come out a reasonable length. (Often, you will want to use multiples of 2, or 5, or 10, etc., to make the axis easy to read.)

√ Now draw in a line for the output axis, look at the (input, output) pairs that you are going to plot, and see what range of outputs to include on this axis. Number this axis, using a spacing that makes it come out a reasonable length. The spacing on the two axes need not be the same.

√ Plot the (input,output) pairs, and sketch a smooth curve from left to right through them.

√ Is this curve the only reasonable possibility, or can you imagine others? If there are others, check the curve by plotting some new (input,output) pairs. In particular, look for other possibilities where the curve bends most sharply, especially around peaks and valleys.

✓ Finally, check the ends of the curve by calculating some (input,output) pairs to the left and right of the ones you already have. If they are too large to plot on your axes, simply imagine where they should go and make sure the curve is headed in that direction.

EXERCISES (Answers start on page 34.)

Sketch graphs of the following rules.

1. $F(x) = (x - 2)(2x - 7)^2$
2. $G(z) = 4z^3 - 27z$
3. $H(a) = 3a^3 + 7a^2 + 4$
4. $V(b) = b(12 - 2b)(10 - 2b)$
5. $K(c) = c(c + 1)(c - 1)$
6. $L(d) = -d(d + 1)(d - 2)$
7. $P(h) = h^2(h - 3)$
8. $R(s) = \frac{1}{3}s^3 - \frac{1}{2}s^2 - 6s + 1$
9. $S(t) = t^2(t - 2)^2$
10. $T(u) = u(u - 1)(u - 2)(u - 3)$
11. $U(v) = (v + 1)(v + 2)(8v^2 + 6v - 11)$
12. $V(w) = w^4$
13. $W(x) = \frac{1}{1 + x^2}$

ANSWERS TO EXERCISES

(1) $F(x) = (x-2)(2x-7)^2$

input	output
~~0~~	~~-98~~
~~1~~	~~-25~~
2	0
3	1
4	2
~~5~~	~~27~~
$2\frac{1}{2}$	2
$3\frac{1}{2}$	0

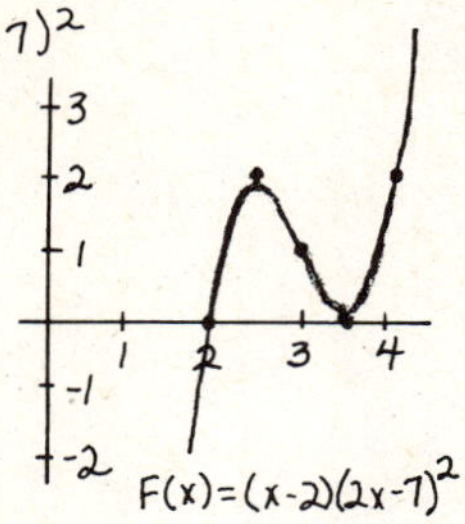

(2) $G(z) = 4z^3 - 27z$

input	output
0	0
1	-23
2	-22
3	27
~~4~~	~~148~~
-1	23
-2	22
-3	-27
~~-4~~	~~-148~~

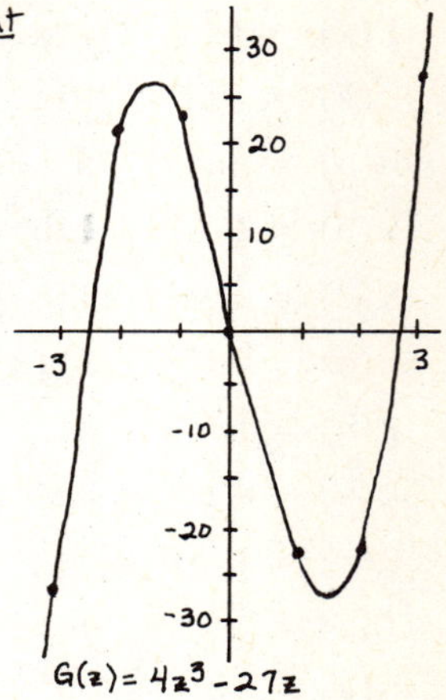

(3) $H(a) = 3a^3 + 7a^2 + 4$

input	output
0	4
1	14
~~2~~	~~56~~
-1	8
-2	8
-3	-14
~~-4~~	~~-76~~

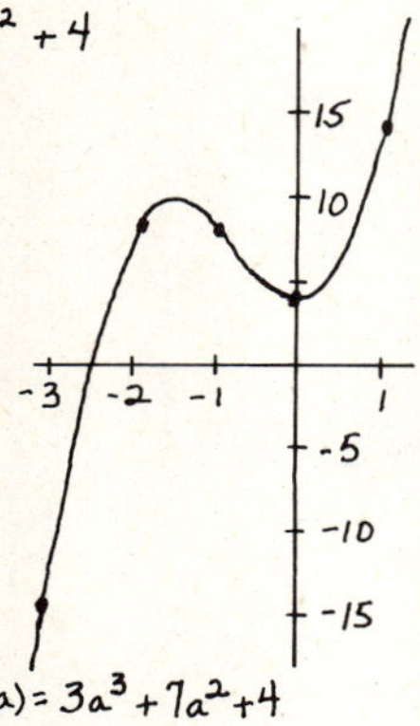

(4) $V(b) = b(12-2b)(10-2b)$

input	output
0	0
1	80
2	96
3	72
4	32
5	0
6	0
7	56
~~8~~	~~192~~
~~-1~~	~~-168~~

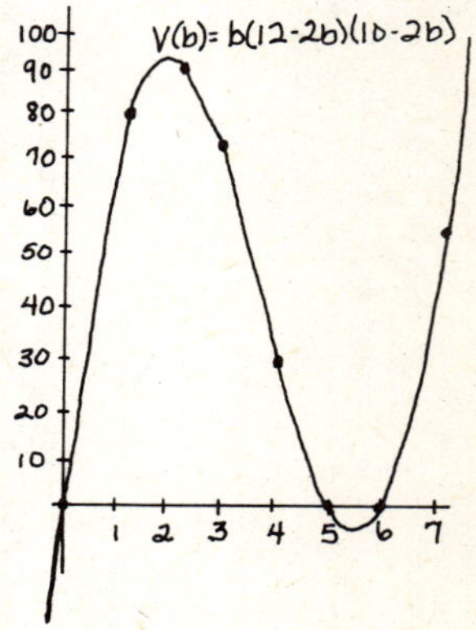

(5) $K(c) = c(c+1)(c-1)$

input	output
0	0
1	0
2	6
~~3~~	~~24~~
-1	0
-2	-6
~~-3~~	~~-24~~

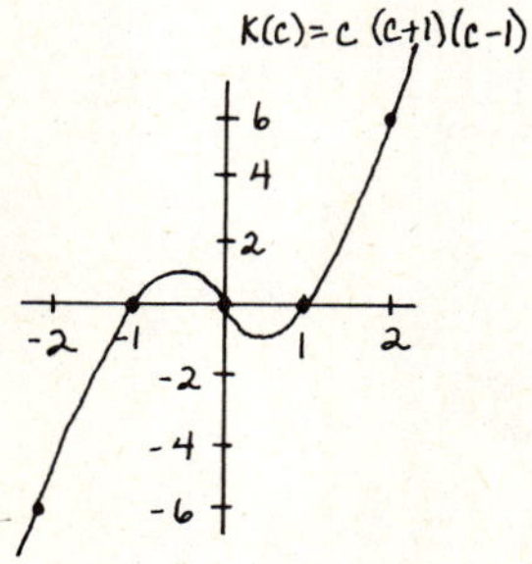

(6) $L(d) = -d(d+1)(d-2)$

input	output
0	0
1	2
2	0
3	-12
~~4~~	~~-40~~
-1	0
-2	8
~~-3~~	~~30~~

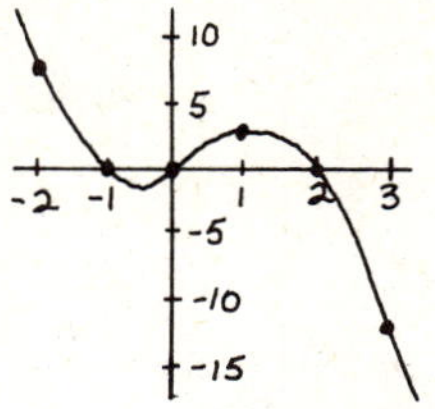

$L(d) = -d(d+1)(d-2)$

(7) $P(h) = h^2(h-3)$

input	output
0	0
1	-2
2	-4
3	0
~~4~~	~~16~~
-1	-4
~~-2~~	~~-20~~

$P(h) = h^2(h-3)$

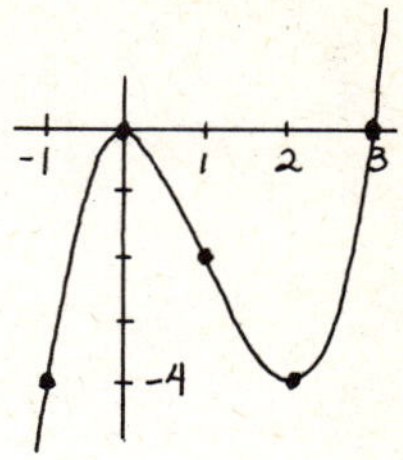

(8) $R(s) = \frac{1}{3}s^3 - \frac{1}{2}s^2 - 6s + 1$

input	output
0	1
1	$-5\frac{1}{6}$
2	$-10\frac{1}{3}$
3	$-12\frac{1}{2}$
4	$-9\frac{2}{3}$
5	$\frac{1}{6}$
~~6~~	~~19~~
-1	$6\frac{1}{6}$
-2	$8\frac{1}{3}$
-3	$5\frac{1}{2}$
-4	$-4\frac{1}{3}$
~~-5~~	~~$-23\frac{1}{6}$~~

$R(s) = \frac{1}{3}s^3 - \frac{1}{2}s^2 - 6s + 1$

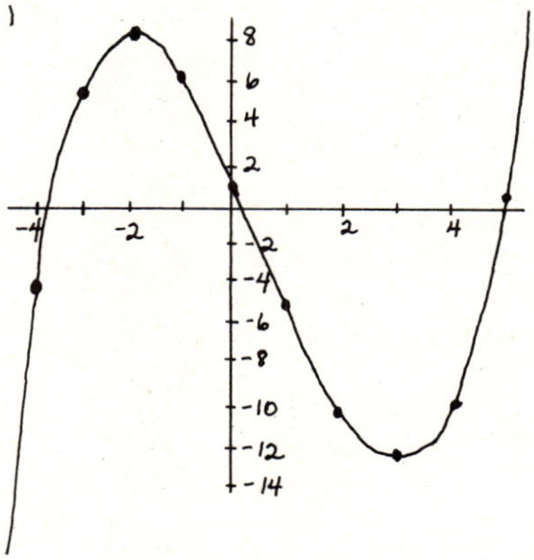

(9) $S(t) = t^2(t-2)^2$

input	output
0	0
1	1
2	0
3	9
~~4~~	~~64~~
-1	9
~~-2~~	~~64~~

$S(t) = t^2(t-2)^2$

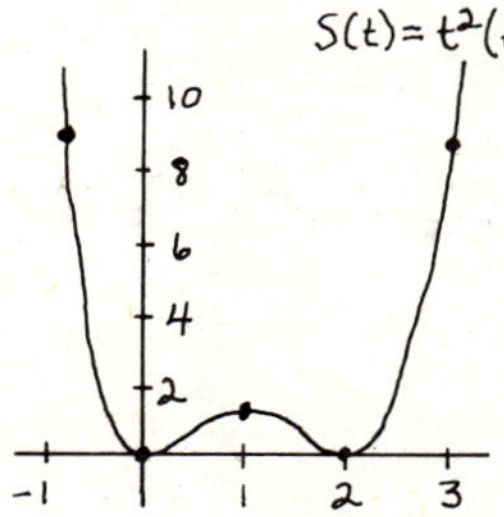

(10) $T(u) = u(u-1)(u-2)(u-3)$

input	output
0	0
1	0
2	0
3	0
4	24
-1	24
$\frac{1}{2}$	$-\frac{15}{16}$
$1\frac{1}{2}$	$\frac{9}{16}$
$2\frac{1}{2}$	$-\frac{15}{16}$

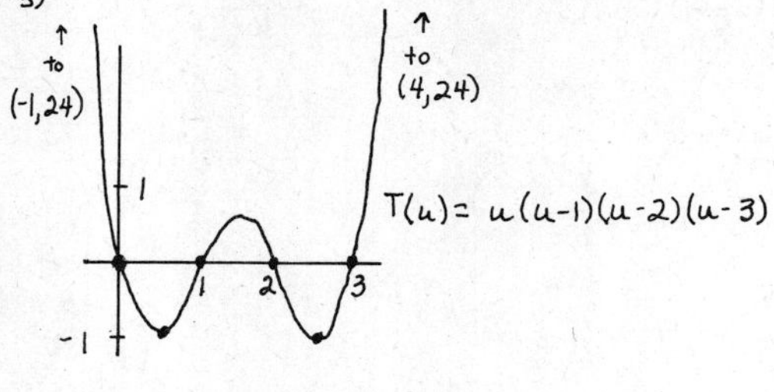

(11) $U(v) = (v+1)(v+2)(8v^2+6v-11)$

input	output
0	-22
1	18
~~2~~	~~396~~
-1	0
-2	0
~~-3~~	~~86~~

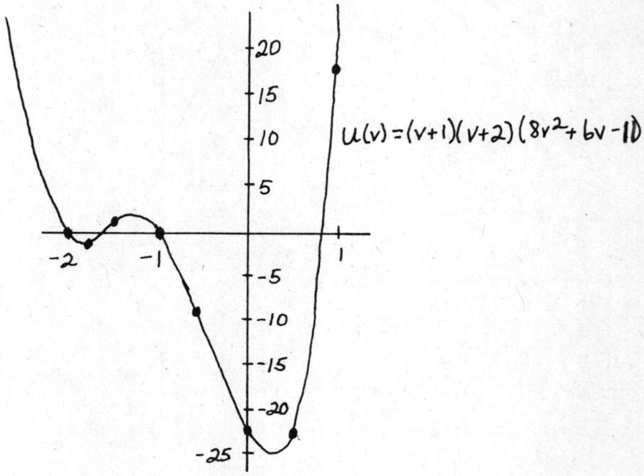

(12) $V(w) = w^4$

input	output
0	0
1	1
2	16
~~3~~	~~81~~
-1	1
-2	16
~~-3~~	~~81~~
$\frac{1}{2}$	$\frac{1}{16}$
$-\frac{1}{2}$	$\frac{1}{16}$

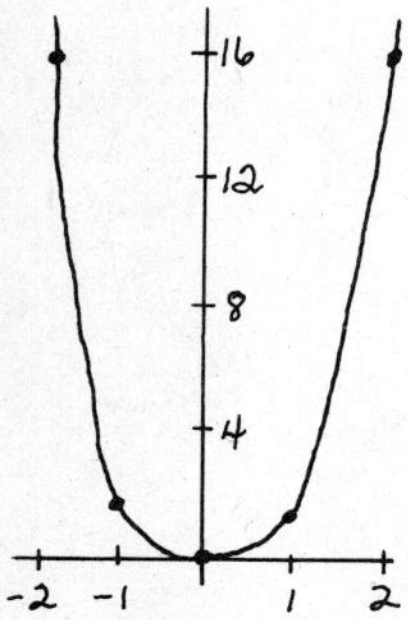

(13) $W(x) = \frac{1}{1+x^2}$

input	output
0	1
1	$\frac{1}{2}$
2	$\frac{1}{5}$
3	$\frac{1}{10}$
-1	$\frac{1}{2}$
-2	$\frac{1}{5}$
-3	$\frac{1}{10}$

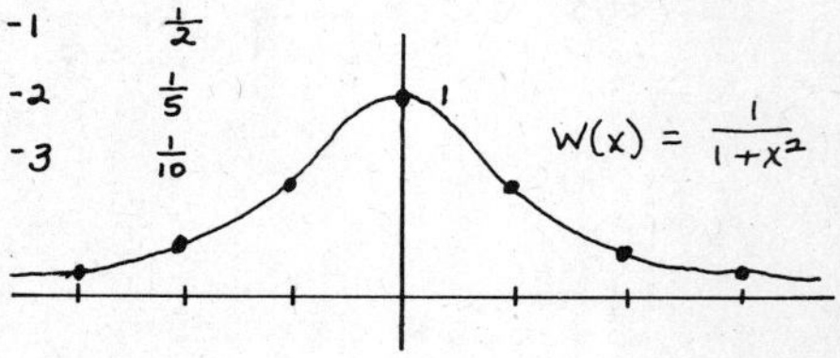

ADDITIONAL EXERCISES

Sketch graphs of the following rules.

1. $F(x) = (x - 3)(2x - 9)^2$
2. $G(z) = 4(z - 1)^3 - 27(z - 1)$
3. $H(a) = 3(a + 1)^3 + 7(a + 1)^2 + 4$
4. $V(b) = (b - 1)(14 - 2b)(12 - 2b)$
5. $K(c) = (c + 1)(c + 2)c$
6. $L(d) = (1 - d)d(d - 3)$
7. $P(h) = (h + 1)^2(h - 2)$
8. $R(s) = \frac{1}{3}(s - 1)^3 - \frac{1}{2}(s - 1)^2 - 6(s - 1) + 1$
9. $S(t) = (t + 1)^2(t - 1)^2$
10. $T(u) = (u - 1)(u - 2)(u - 3)(u - 4)$
11. $U(v) = (v + 2)(v + 3)(8v^2 + 22v + 3)$
12. $V(w) = (w + 1)^4$
13. $W(x) = \dfrac{1}{1 + (x - 1)^2}$

SUMMARY OF LESSON 2

This lesson makes two important points: First, we should check our graphs carefully where they bend most sharply, especially around the peaks and valleys of the curve. Second, the spacing between the numbers on the input axis need not be the same as the spacing on the output-axis. Sometimes there are advantages to using a different spacing. These points are developed throughout the lesson, and are summarized on pages 26 and 31-32. (Caution: don't use different spacings on the same axis. For instance, don't draw an x-axis like this: 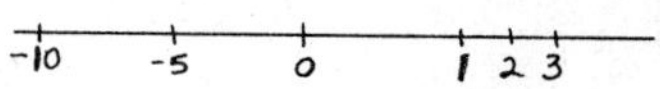

WRONG!)

Example 2 introduces the graph of a 4th degree formula (where the highest power of x is x^4). These are w- or u-shaped curves, such as:

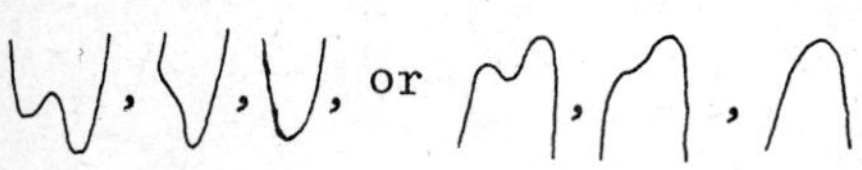

.

If you have trouble with any of the exercises, and the answers don't give enough help, then reread the lesson carefully. There are only two examples in the lesson, and each one is relevant to each exercise.

Lesson 3

GRAPH REASONING I

FINDING ZEROS BY GRAPHING

EXAMPLE 1 (Review) rule F: $F(x) = x^2 + x - 12$

What inputs to rule F give an output of 0?

We know two ways to answer this question. One way is to write an equation and solve for the unknown input.

$$x^2 + x - 12 = 0$$

$$(x + 4)(x - 3) = 0$$

-4, 3 are the solutions.

So if we use -4 or 3 as inputs to rule F, we will get 0 as an output. (Check this.) This was covered in Module X, lesson 8, example 1.

We can also answer the question by looking at the graph of rule F.

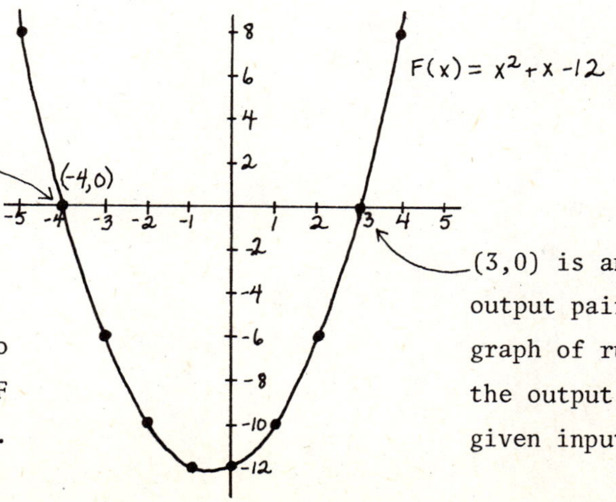

(-4,0) is an input-output pair on the graph of rule F. So the output of rule F given input -4 is 0.

(3,0) is an input-output pair on the graph of rule F. So the output of rule F given input 3 is 0.

Answer: -4 and 3

This was covered in Module V, lesson 6.

In this example, it was easier to solve the equation than to draw the graph. But as we shall soon see, finding inputs that give an output of 0 is sometimes more easily done by graphing.

Note: The basic idea is that the y-coordinate of every point on the x-axis is 0.

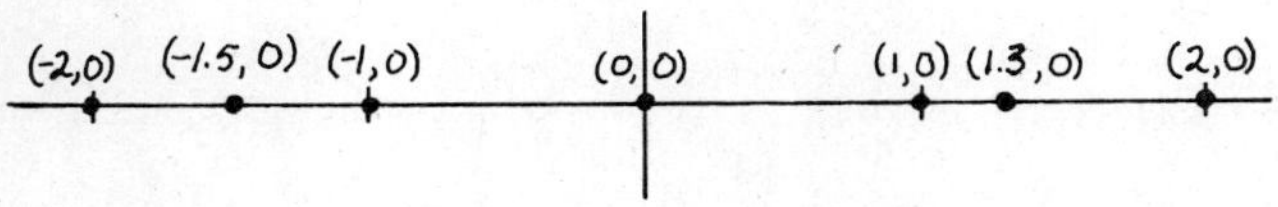

So wherever a graph touches the x-axis, the rule has an output of 0.

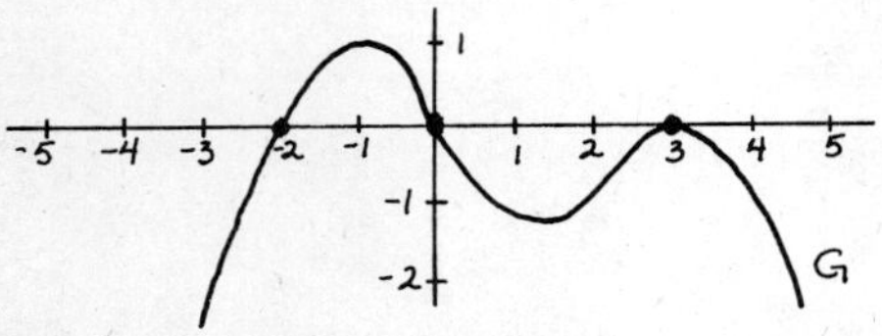

This is the graph of a rule that gives an output of 0 for inputs of -2, 0, and 3.

There is some special mathematical vocabulary for this situation. If an input to rule F gives an output of 0, then the input is called a zero of rule F.

For instance, in example 1 we used the rule $F(x) = x^2 + x - 12$. We saw that the zeros of this rule were -4 and 3.

The graph of rule G above shows that the zeros of rule G are -2, 0. and 3.

EXAMPLE 2 rule F: $F(x) = 3x^3 + 7x^2 - 4$

rule G: $G(x) = 3x^3 + 7x^2$

rule H: $H(x) = 3x^3 + 7x^2 + 4$

Find the zeros of each rule by graphing:

rule F

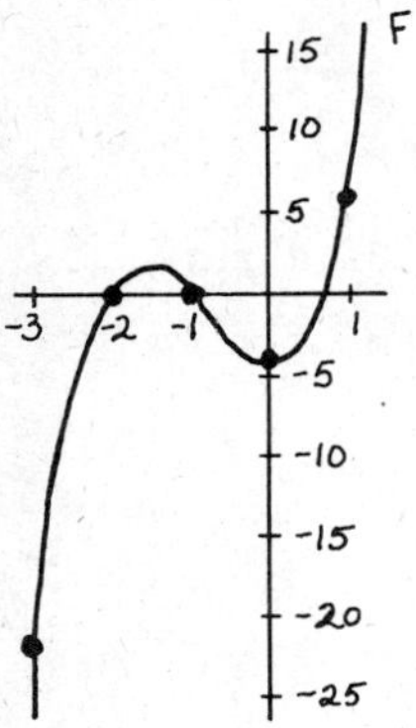

Rule R has three zeros. They are -2, -1, and a number between 0 and 1.

rule G

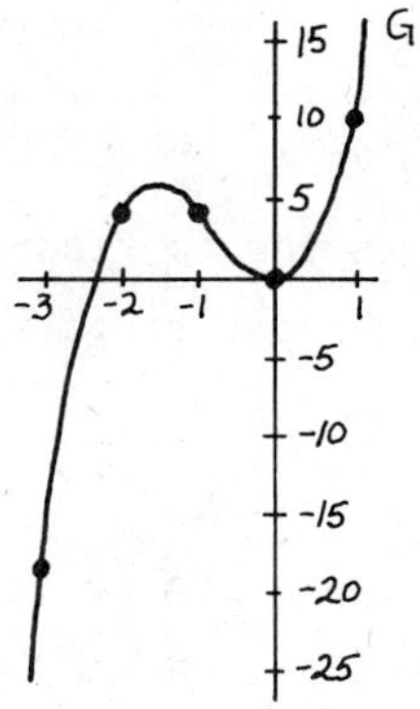

Rule G has two zeros. One of them is a number between -3 and -2, and the other one is 0.

rule H

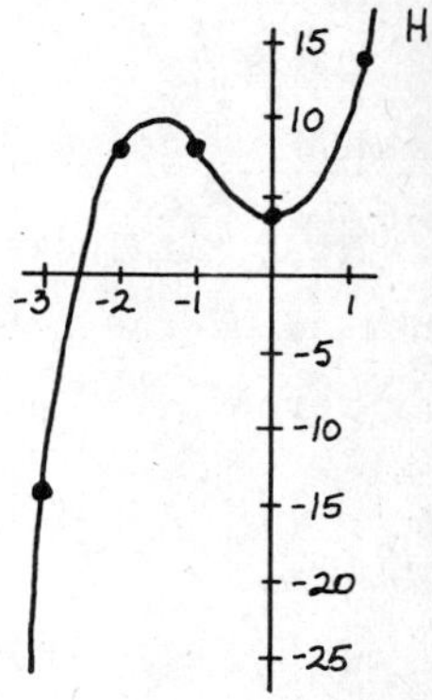

Rule H has one zero. It is a number between -3 and -2.

Notice that some of these zeros were at points that we had plotted, and in these cases we could give their values exactly. But in the other cases, the zeros were between points we had plotted, and we had to give estimates.

Actually, these graphs were drawn carefully, and by reading them accurately we could get closer estimates (-2,-1, and 0.7 for F; -2.3 and 0 for G; and -2.6 for H). But for now, reading graphs that precisely is not important. The important point is this:

Wherever a graph touches or crosses the x-axis, the rule has an output of zero. So to find the zeros of a rule, you can draw the graph and see where it touches or crosses the x-axis. You can tell how many zeros there are, and you can estimate them roughly. (Between-whole-number estimates are good enough in this module.)

We can also use graphing to find inputs that give outputs other than zero.

<u>EXAMPLE 3</u> Solve x(12 - 2x)(10 - 2x) = 50 by graphing.

At first, it looks like we are trying to find the "fifties" of a rule, instead of the zeros. But we can convert this into a problem of finding zeros.

$$\begin{aligned} x(12 - 2x)(10 - 2x) &= 50 \\ -50 \quad &\quad -50 \\ \hline x(12 - 2x)(10 - 2x) - 50 &= 0 \end{aligned}$$

So we can graph F(x) = x(12 - 2x)(10 - 2x) - 50, and find its zeros. These will be the solutions to the original equation.

F(x) = x(12 - 2x)(10 - 2x) - 50

(input,output)
(-1,-218)
(0,-50)
(1,30)
(2,46)
(3,22)
(4,-18)
(5,-50)
(6,-50)
(7,6)
(8,142)

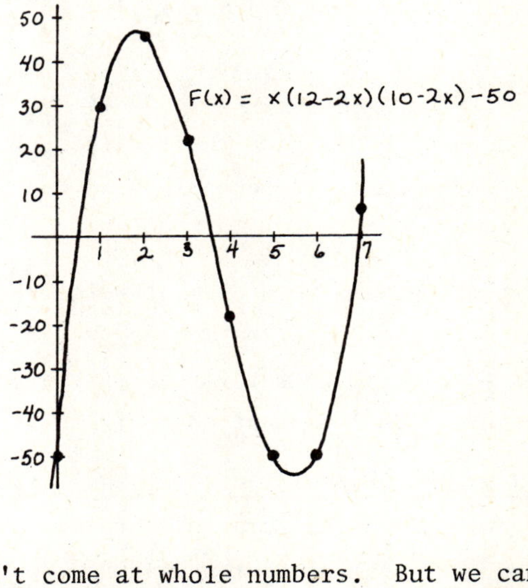

Unfortunately, the zeros didn't come at whole numbers. But we can see that there are three solutions. One is between 0 and 1, one is between 3 and 4, and one is between 6 and 7. (You might try to read the graph more accurately. Then you could put your estimates into the original equation, and see how close the output is to 50. But for the exercises in this module, between-whole-number estimates are good enough.)

Let's see how these techniques can be useful in a realistic situation.

EXAMPLE 4 (This repeats example 11 of lesson 3, Module X.) Suppose you have some sheets of cardboard that are 12 inches long and 10 inches wide, and you want to make some boxes by cutting little squares out of the corners and then folding up the resulting "tabs" on each side.

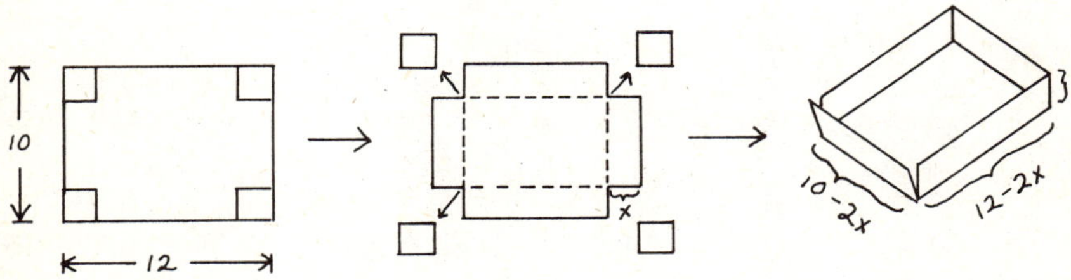

How long should the cuts be (that is, how long should the side of each little square be) so that the volume of the resulting box will be 50 cubic inches?

Pages 42-43 in Module X showed how to write an input-output rule for this situation. (It would be a good review to do this for yourself before reading on.) The formula for this rule was

$$V(x) = x(12 - 2x)(10 - 2x).$$

The cut-length is the input, and the volume of the resulting box is the output.

In order to solve this word problem, we have to find an input to rule V that gives an output of 50. That is, we have to solve

$$x(12 - 2x)(10 - 2x) = 50.$$

It would be very difficult to solve this equation algebraically, but we saw in the last example that we can get approximate solutions quite easily by graphing. Do all three of those solutions work here?

The solutions between 0 and 1 and between 3 and 4 are OK. We can certainly cut out little squares of that size. But we can't cut squares out of each corner that are between 6 and 7 inches long. The whole sheet is only 10 inches wide, so 5-inch squares are the largest we could cut out of each corner.

So there are two different ways to make a 50-cubic-inch box by the method described in the problem. One way involves cutting small squares (between 0 and 1 inches long) out of each corner. The other way involves cutting fairly large squares (between 3 and 4 inches long) out of each corner. Both ways give the same 50-cubic-inch volume. (To find out more exactly what lengths to use, we would read the graph more accurately.)

FINDING MAXIMUMS AND MINIMUMS BY GRAPHING

EXAMPLE 5 By making cardboard boxes as in example 4, what is the largest volume we can get?

Here is a new kind of problem. We don't have a specific input or output to work with, so we can't simply write an equation and solve it.

Let's work out some samples, and see if we can get some idea of what the largest volume must be. Rather than figure out the volume each time from the original description, it will be easier to use the input-output rule that we found:

$$V(x) = x(12 - 2x)(10 - 2x)$$

input (cut-length)	output (volume)
0	0
1	80
2	96 ← largest volume?
3	72
4	32
5	0

There is no point in trying inputs less than 0, since a negative cut-length doesn't make sense. And as we saw in example 4, we can't make cuts longer than 5 inches, since each sheet is only 10 inches wide.

Does this mean that 96 cubic inches is the largest possible volume? Not necessarily, since some input between the whole numbers we have tried might give a larger output. One way to find out is to plot these input-output pairs, and draw the graph of rule V.

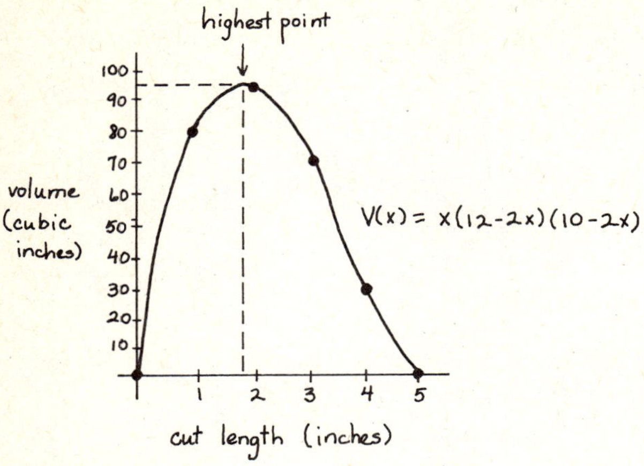

<u>The highest point on the graph shows the maximum output</u>. The way we have sketched it, the graph reaches its peak where the output is a little larger than the 96 that we plotted. So according to this graph, the maximum volume is between 96 and 100 cubic inches.

We could get a more precise answer by checking the curve carefully around its peak, and reading the graph more accurately. But again, extreme accuracy in graphing is not important in this module. We are mainly interested in learning to sketch graphs and interpret them. Different people might sketch the curve above in slightly different ways, and get slightly different estimates for the answer. For now, the important thing is to know the basic ideas involved in finding maximums:

If a word problem asks you to find the maximum of some quantity:

- ✓ Write a rule that gives that quantity as output.
- ✓ Graph the rule, using inputs that make sense in the realistic situation.
- ✓ Identify the highest point on the graph.
- ✓ Estimate the maximum output (according to <u>your</u> graph).

Knowing how to follow the first three steps is the most important part. As long as you can do this, don't worry about the precision of your answer. In calculus, you will learn an algebraic technique to find such answers exactly, and then you can use this graphing technique as a check.

<u>Note</u>: In this example, it was important that we didn't use inputs larger than 5, since these didn't make sense in the realistic situation. Otherwise, we might have drawn the following graph:

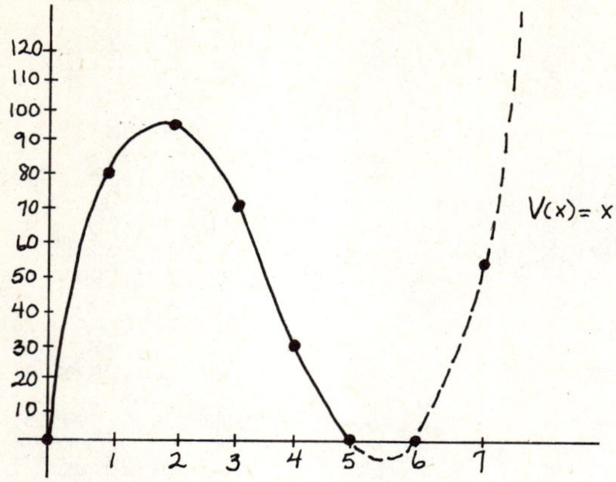

The dashed part of this curve shows some outputs much larger than the solid part. In fact, the outputs keep getting larger and larger—there is no limit. But we have to remember that the dashed part of the curve doesn't apply to the realistic situation (since we can't actually make cuts longer than 5 inches). <u>When you are using an input-output rule to describe a realistic situation, it is important to use only inputs that make sense.</u>

<u>EXAMPLE 6</u> A bus carries an average of 720 passengers each day from Berkeley to San Francisco. The fare is 60 cents. The company finds that 100 more people will ride each day for each 5-cent reduction in the fare. What fare should the company charge in order to make the most money?

We will write out a rule that takes the fare as input and gives the total amount of money collected each day as output. Then we will graph the rule and find the highest point. This will tell us what fare gives the maximum income.

We did the first part of this on pages 22-24 and 40-41 of Module X. We chose some sample fares and calculated the outputs. Then we used these calculations as a guide to write out the rule in English. Finally, we translated the English into a formula. (Review this if you have any trouble doing it now.)

$$B(x) = \left(\frac{60 - x}{5} \cdot 100 + 720\right)x$$

This rule takes the fare (in cents) as input, and gives the total amount of money collected each day as output.

Now we want to draw the graph and find the highest point. What inputs should we use? One way to cut down on the number of points that we have to plot is to notice that the formula is a quadratic.

$$B(x) = \left(\frac{60 - x}{\cancel{5}\,_1} \cdot \overset{20}{\cancel{100}} + 720\right)x = [(60 - x)20 + 720]x$$

$$= [1200 - 20x + 720]x = (1920 - 20x)x = 1920x - 20x^2$$

So the graph will be a parabola, and will have only one peak (or valley). We can decide approximately where this peak is from a table of inputs and outputs. Then we can draw a graph that uses only inputs in this area.

$$B(x) = 1920x - 20x^2$$

First let's factor to make the arithmetic easier.

$$B(x) = (96 - x)20x$$

input (in cents)	output (in cents)
0	0
10	17,200
20	30,400
30	39,600
40	44,800
50	46,000
60	43,200

(Outputs 40–60:) The maximum output is somewhere in this range.

It looks as though the peak is somewhere between inputs of 40 and 60, so we will draw a graph for these inputs:

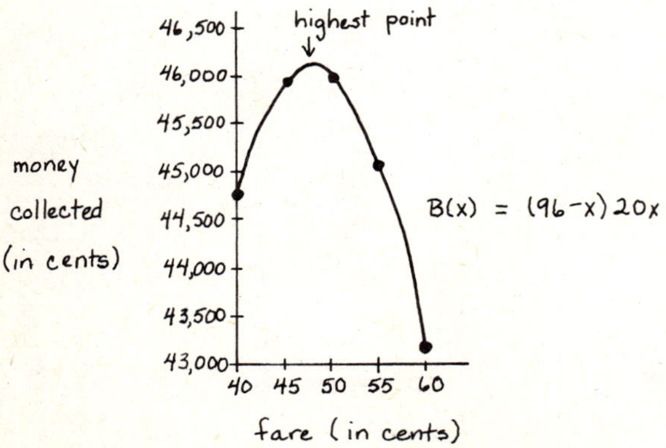

Answer: The fare that results in the highest daily income is somewhere between 45 and 50 cents. (Notice that the problem didn't ask for the maximum output. It asked for the input that gives the maximum output. It asked for the fare that gives the largest income.)

In this example, we could have found the answer algebraically. We learned in Module V, lesson 6, that the input for the top point on a parabola is halfway between the zeros (if there are any). But this method does not work in general, only for parabolas. For instance, the graph on page 45 is the graph of a cubic. You can see that the input for the top point is not halfway between the zeros. In this module, we will continue to sketch the graph and look for the highest point.

EXAMPLE 7 rule F: $F(t) = t(t + 1)(8t^2 - 10t - 9)$
Find the zeros and the minimum and maximum outputs.

This rule was graphed in lesson 2, example 2. This time we are not dealing with a realistic situation, so we don't have to worry about any restrictions on the inputs. Also, this time we are looking for the minimum output, as well as the maximum, so we will look for the lowest, as well as the highest, points. The graph from lesson 2 is copied below.

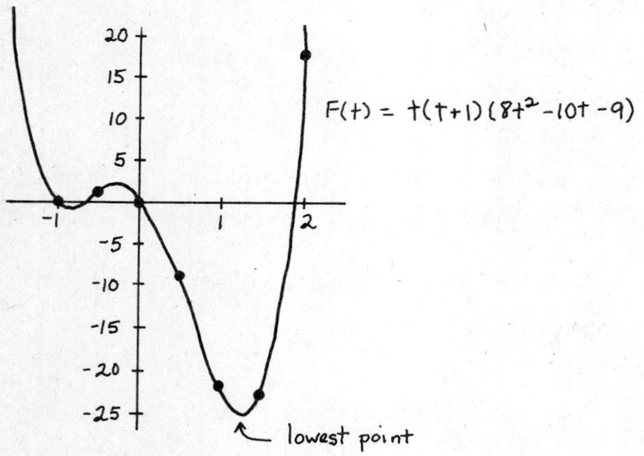

Answers: Rule F has 4 zeros: -1, a number between -1 and 0, 0, and a number between 1 and 2. The minimum output is about -25. There is no maximum output. (Notice that there is a relatively high point between -1 and 0. But there is no overall highest point— the outputs just get larger and larger without limit.)

EXAMPLE 8 rule F: $F(t) = t(t + 1)(8t^2 - 10t - 9)$, with inputs restricted to numbers from -1 to 1. Find the minimum and maximum outputs in this situation.

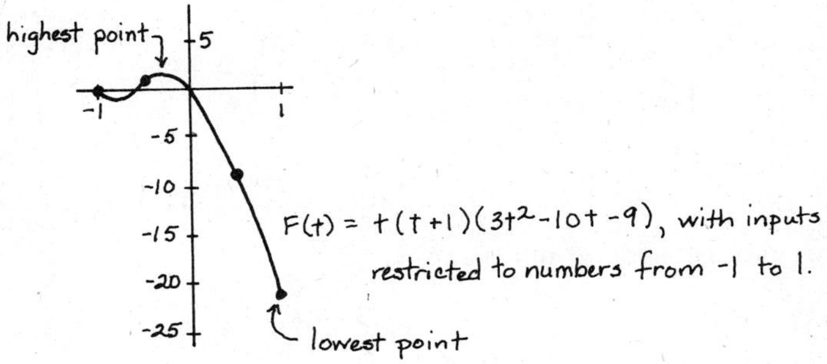

The graph above is all that is left of the graph in example 7 when only the allowed inputs are used. In this case, the high point between -1 and 0 is the highest point. It shows a maximum output somewhere between 0 and 5. The minimum output is -22. We can calculate this value exactly, since we know that 1 is the corresponding input.

EXERCISES (Answers start on page 51.)

Find the zeros of the following rules. If the zeros are not whole numbers, then give between-whole-number estimates.

1. $F(x) = x^3 - 4x$
2. $G(z) = z^3 - 4z + 1$
3. $H(r) = r^3 - 4r - 1$

Solve the following equations. If the solutions are not whole numbers, then give between-whole-number estimates.

4. $3x + 2x^2 - x^3 = 0$
5. $3x + 2x^2 - x^3 = 4$
6. $3x + 2x^2 - x^3 = -2$

Find (approximately) the maximum and minimum outputs and the corresponding inputs. Show the high and low points on the graph.

7. $F(x) = x^2 - 5x + 8$
8. $G(x) = -x(x + 1)(x - 1)(x - 3)$
9. $H(x) = x^2(3x^2 - 4x - 12)$
10. $F(x) = x^3 - 5x$, with inputs restricted to numbers from -2 to 2
11. $G(x) = x^3 - 5x$, with inputs restricted to numbers from -3 to 3
12. A peach tree will produce an average of 200 pounds of fruit each year, unless it is too crowded by other trees. Fifty trees per acre is alright, but each additional tree over 50 will reduce the yield of each tree on the acre by 1 pound. How many trees per acre should you plant so that the total yield of the acre will be as large as possible? (Show the high point on your graph.)
13. You have to fence in a rectangular play area on the lawn of a day care center. You can choose the length of the rectangle, but you only have 200 feet of fence to work with. What is the maximum area you can enclose with this amount of fence? (Show the high point on your graph.)
14. Suppose that you can run at a steady speed of 8 miles per hour for 1 hour. But if you slow down, you can run longer. In fact, for every mile per hour slower that you run, you can run for an extra hour. What speed should you run in order to go as far as possible? (Show the high point on your graph.)
15. Suppose that you are building a box with a square base. The length and the height must add up to 3 feet. What is the maximum volume the box can have? What length for the base will give this volume? (Show the high point on your graph.)

ANSWERS TO EXERCISES

(1)

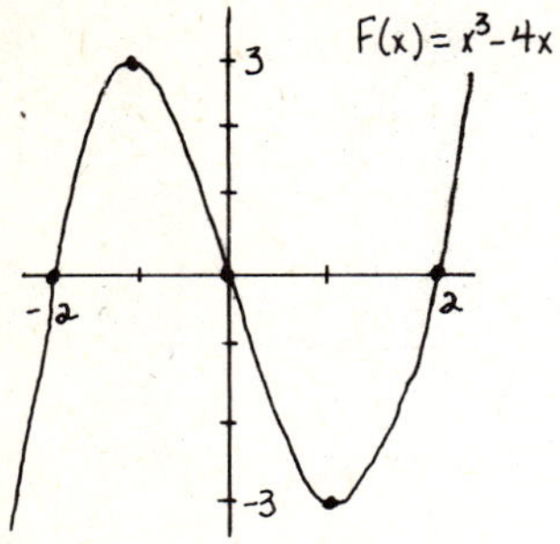

The zeros of F are -2, 0, and 2.

(2)

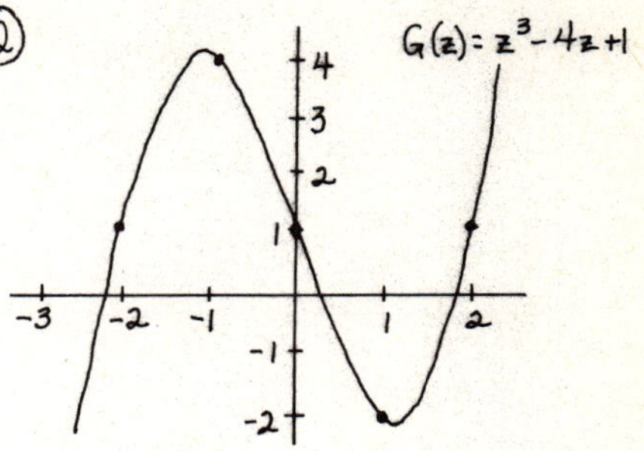

The zeros of G are
✓ a number between -2 and -3,
✓ a number between 0 and 1,
✓ a number between 1 and 2.

(3)

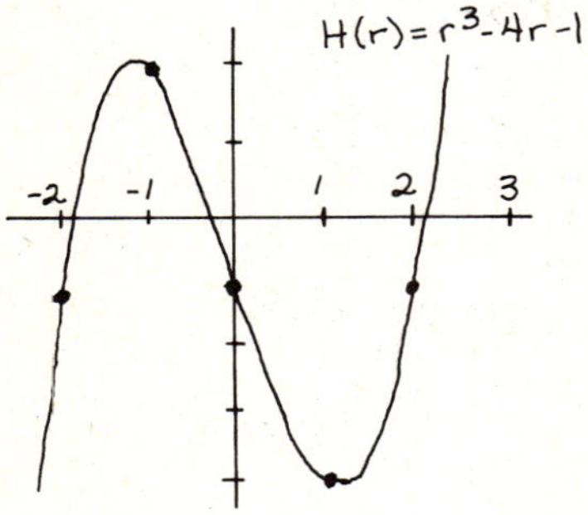

The zeros of H are
✓ a number between -2 and -1,
✓ a number between -1 and 0,
✓ a number between 2 and 3.

(4)

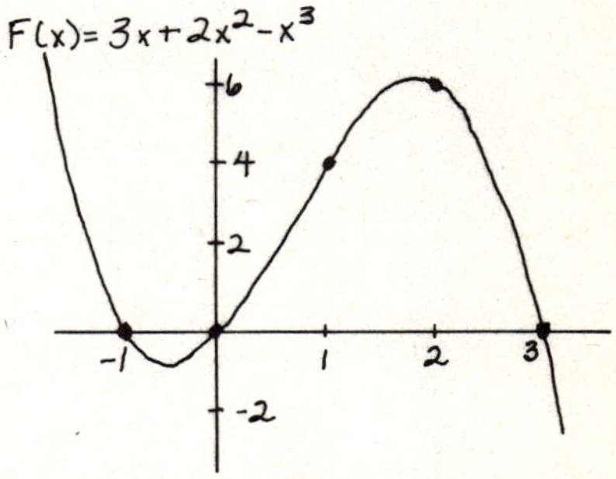

The solutions of $3x+2x^2-x^3=0$ are -1, 0, and 3.

(5)

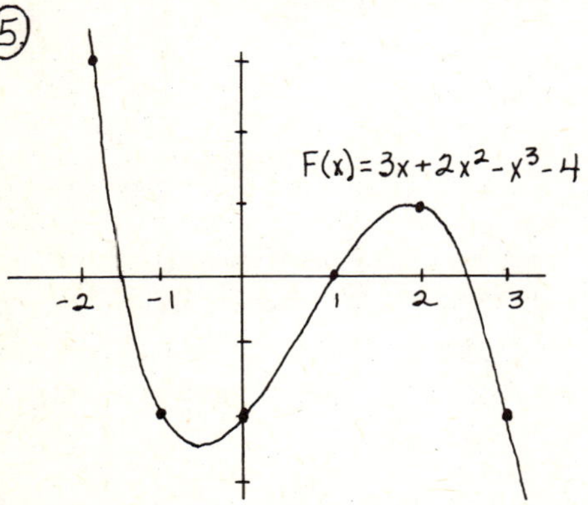

The solutions of $3x+2x^2-x^3=4$ are
✓ a number between -2 and -1,
✓ 1,
✓ a number between 2 and 3.

(6)

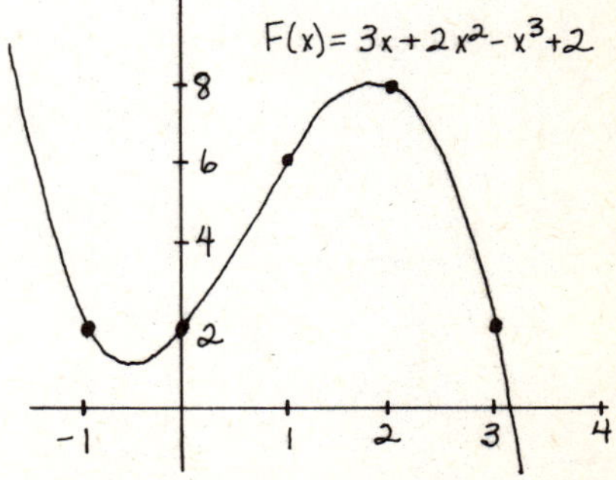

The only solution to $3x+2x^2-x^3=-2$ is a number between 3 and 4.

(7)

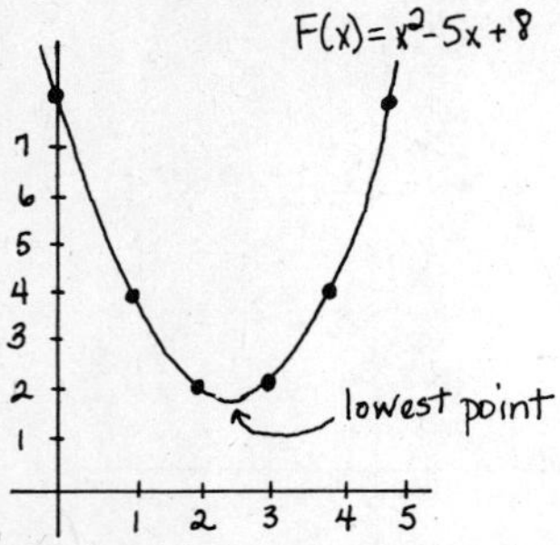

The minimum output is between 1 and 2.
The corresponding input is between 2 and 3.
There is no maximum output.

(8)

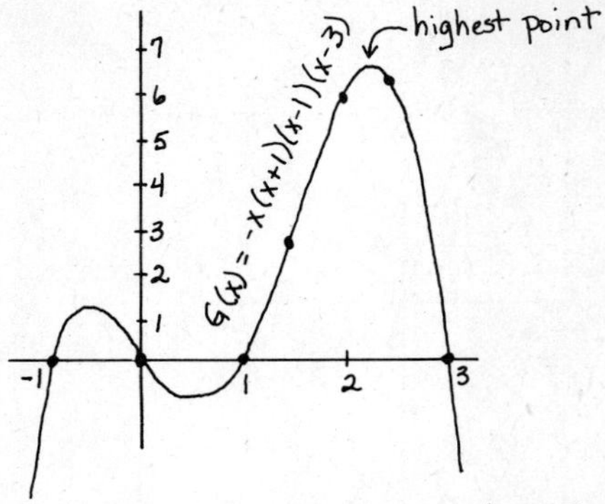

The maximum output is between 6 and 7.
The corresponding input is between 2 and 3.
There is no minimum output.

(9)

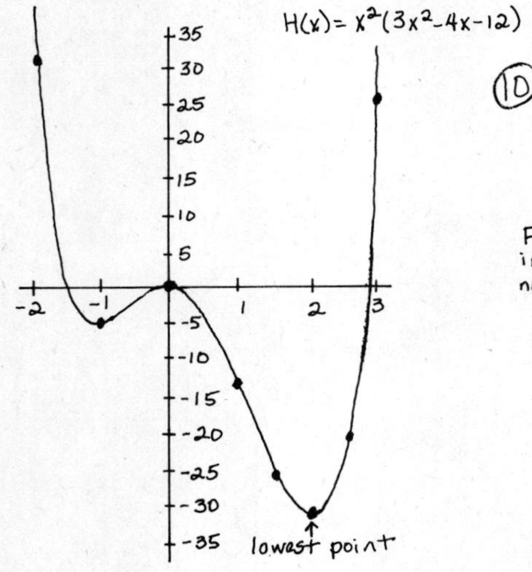

The minimum output is about -32.
The corresponding input is about 2.
There is no maximum output

(10)

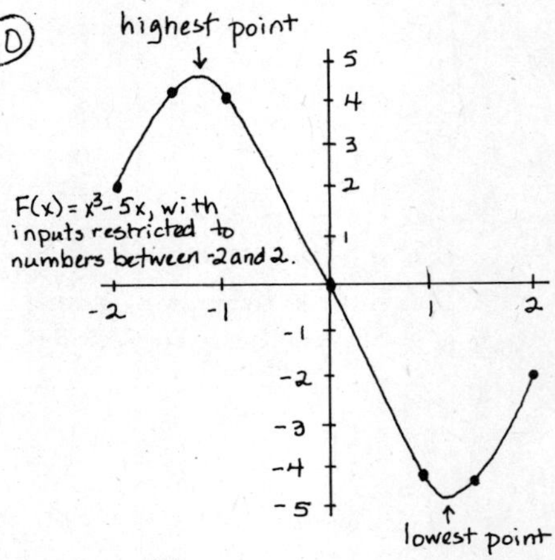

The maximum output is between 4 and 5.
The corresponding input is between -2 and -1.
The minimum output is between -4 and -5.
The corresponding input is between 1 and 2.

(11)

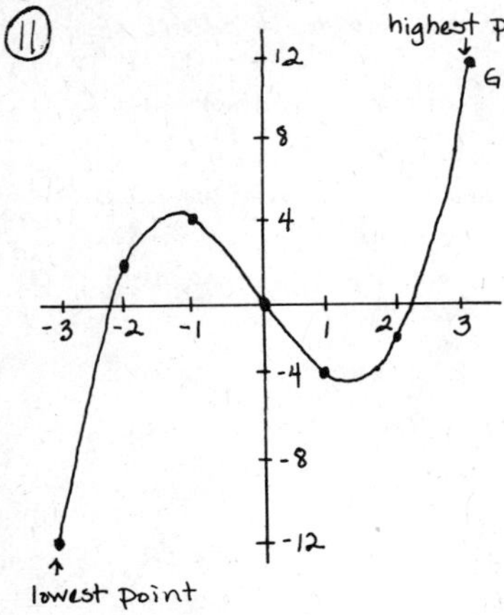

$G(x) = x^3 - 5x$, with inputs restricted to numbers between -3 and 3

The minimum output is -12.
The corresponding input is - 3.
The maximum output is 12.
The corresponding input is 3.

(12) (The first part of this problem was covered in detail in Module I, on pages 20-21 and 38 - 39)

First, we need a rule that takes the number of trees per acre as input, and gives the total yield of the acre as output.

- Try 60 trees per acre. This is 10 over the limit, so the yield of each tree will drop to 190 pounds. So there are 60 trees on the acre, each producing 190 pounds, giving 60 • 190 pounds in all. This is 11,400 pounds.
- Try 65 trees per acre. This is 15 over the limit, so the yield of each tree will drop by 15 pounds, down to 185. 65 trees, each producing 185 pounds, will yield 65 • 185 pounds in all. This is 12, 025 pounds.

First sample	Second sample	Rule P	Formula
(60) - 50 = 10	(65) - 50 = 15	Take the input and subtract 50.	$x - 50$
200 - 10 = 190	200 - 15 = 185	Subtract this from 200.	$200 - (x-50)$
190 • (60) = 11,400	185 • (65) = 12,025	Multiply this by the input.	$[200-(x-50)] \cdot x$
		This is the output.	$P(x) = [200-(x-50)] \cdot x$

$P(x) = [200 - (x-50)] \cdot x = [200 - x + 50]x = [250-x]x$

Now we want to graph $P(x) = 250x - x^2$ and find the highest point.

input	output
60	11,400
80	13,600
100	15, 000
120	15, 600
140	15, 400
160	14, 400

(100–140:) The maximum output is somewhere in this range.

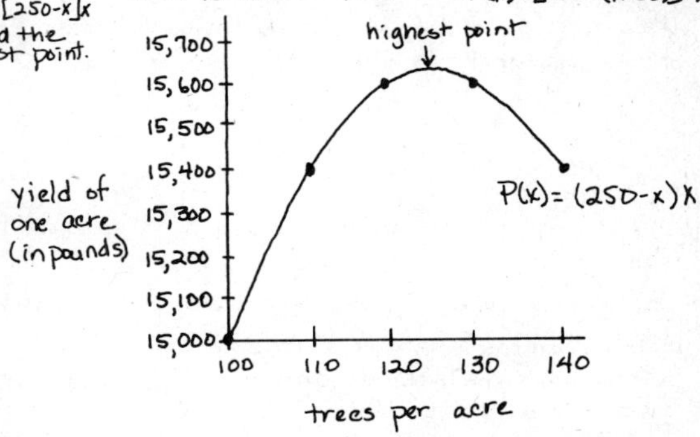

The question was "How many trees per acre should you plant so that the total yield of the acre will be as large as possible?"

The answer is somewhere between 120 and 130 trees per acre.
(A more accurate graph would give a more precise answer.)

(13) (The first part of this problem was covered in detail in Module I, on pages 24- 26 and page 44.)

First, we need a rule that takes the length of the rectangle as input, and gives its area as output.

- Try a length of 80. Then 2 sides of the rectangle are 80 feet long, using up 160 feet of fence. This leaves 40 feet (out of the 200) for the other two sides. So each of these must be 20 feet long. So the rectangle is 80 feet long and 20 feet wide. Its area is $20 \cdot 80 = 1600$ square feet.

Try a length of 60. 2 sides, 60 feet each, make 120 feet. This leaves 80 feet for the other sides, so they must be 40 feet each. $40 \cdot 60 = 2400$

First sample	Second sample	Rule	Formula
$(80) \cdot 2 = 160$	$(60) \cdot 2 = 120$	Multiply the input by 2.	$2x$
$200 - 160 = 40$	$200 - 120 = 80$	Subtract this from 200.	$200 - 2x$
$\frac{40}{2} = 20$	$\frac{80}{2} = 40$	Divide the result by 2.	$\frac{200-2x}{2}$
$20 \cdot (80) = 1600$	$40 \cdot (60) = 2400$	Multiply this by the input. This is the output.	$\left(\frac{200-2x}{2}\right)x$

$$A(x) = \left(\frac{200-2x}{2}\right)x$$

$$A(x) = \left(\frac{200-2x}{2}\right)x = \left(\frac{\cancel{2}(100-x)}{\cancel{2}}\right)x = (100-x)x$$

input	output
30	2,100
40	2,400
50	2,500
60	2,400
70	2,100

The maximum area is in this range (40–60).

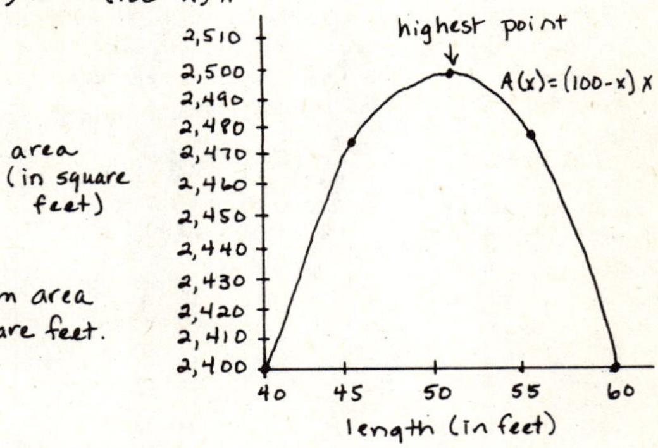

The graph shows that the maximum area of the rectangle is about 2500 square feet.

(14) (The first part of this problem was covered in Module I on page 46, exercise 19, and page 50, answer to exercise 19.)

We need a rule that takes running speed as input, and gives total distance as output.

- Try 6 miles per hour. This is 2 miles per hour slower, so you can run for an extra 2 hours, for a total of 3 hours. Running for 3 hours at 6 miles per hour will take you 18 miles.
- Try 5 miles per hour. This is 3 miles per hour slower, so you can run for an extra 3 hours, for a total of 4 hours. Running for 4 hours at 5 miles per hour will take you 20 miles.

First sample	Second sample	Rule D	Formula
8 - (6) = 2	8 - (5) = 3	Subtract the input from 8.	8 - x
2 + 1 = 3	3 + 1 = 4	Add 1 to this, and	8 - x + 1
3 · (6) = 18	4 · (5) = 20	multiply by the input.	(8 - x + 1) x
		This is the output.	D(x) = (8 - x + 1) x

$D(x) = (8 - x + 1)x = (9 - x)x$

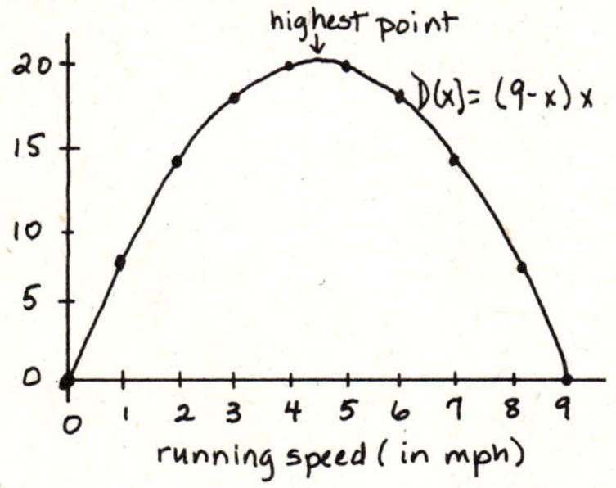

So in order to run as far as possible, you should run between 4 and 5 miles per hour.

(15) First, we need a rule that takes the length of the base as input and gives the volume as output.

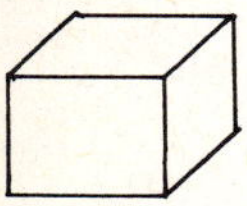

Try a base length of 1 foot. Then the height must be 2 feet. The volume will be 2·1·1 = 2 cubic feet.

Try a base length of 2 feet. Then the height must be 1 foot. The volume will be 1·2·2 = 4 cubic feet.

Sample 1	Sample 2	Rule V	Formula
3 - (1) = 2	3 - (2) = 1	Subtract the input from 3.	3 - x
2·(1)·(1) = 2	1·(2)·(2) = 4	multiply this number times the input times the input.	(3-x)·x·x
		This is the output.	V(x) = (3-x)·x·x

$V(x) = (3-x)\cdot x^2$ In graphing this, we should not use negative inputs, since the base length can't be negative. And since the base + height = 3, the base length can't be more than 3. (The height can't be negative.)

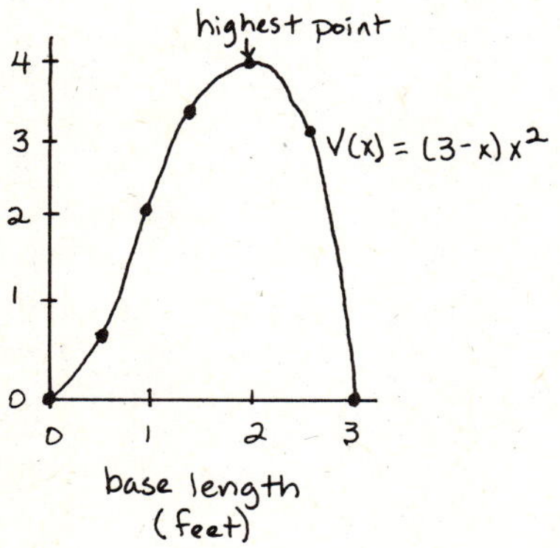

The maximum volume is about 4 cubic feet. The corresponding base length is about 2 feet.

ADDITIONAL EXERCISES

Find the zeros of the following rules. If the zeros are not whole numbers, then give between-whole-number estimates.

1. $F(x) = (x - 1)^3 - 4x$
2. $G(z) = (z - 1)^3 - 4(z - 1) + 1$
3. $H(r) = (r - 1)^3 - 4(r - 1) - 1$

Solve the following equations. If the solutions are not whole numbers, then give between-whole-number estimates.

4. $3(x + 1) + 2(x + 1)^2 - (x + 1)^3 = 0$
5. $3(x + 1) + 2(x + 1)^2 - (x + 1)^3 = 4$
6. $3(x + 1) + 2(x + 1)^2 - (x + 1)^3 = -2$

Find (approximately) the maximum and minimum outputs and the corresponding inputs. Show the high and low points on the graph.

7. $F(x) = (x - 1)^2 - 5(x - 1) + 8$
8. $G(x) = (1 - x)x(x - 2)(x - 4)$
9. $H(x) = (x - 1)^2(3x^2 - 10x - 5)$
10. $F(x) = x^3 + 3x^2 - 2x - 4$, with inputs restricted to numbers from -3 to 1
11. $G(x) = x^3 + 3x^2 - 2x - 4$, with inputs restricted to numbers from -4 to 2
12. Suppose that uncrowded peach trees yield 180 pounds of fruit apiece per year, but that every additional tree over the limit of 60 per acre will reduce the yield of each tree by 2 pounds. How many trees per acre should you plant so that the total yield of the acre will be as large as possible? (Show the high point on your graph.)
13. You want to fence in a pasture next to a river. The fence will be rectangular, and you want to leave the river side unfenced to allow the cattle to drink. You can choose the pasture's length along the river, but you only have 600 feet of fence to work with. What length (along the stream) should the pasture be in order to have the maximum area? (Show the high point on your graph.)
14. You can peddle your bicycle at 12 miles per hour for 4 hours straight before you quit from exhaustion. But for every mile per hour slower that you go, you can peddle for an extra 2 hours. At what speed should you peddle in order to go as far as possible? (Show the high point on your graph.)

15. Suppose that you are building a box with a square base. The length and height must add up to 6 feet. What is the maximum volume the box can have? What length for the base will give this volume? (Show the high point on your graph.)

SUMMARY OF LESSON 3

The graph of a rule can give us a lot of information about the rule that might otherwise be hard to find. We can find the inputs to a rule that give an output of zero [pages 39-44], and we can find the maximum or minimum output of a rule [pages 44-49].

If an input to rule F gives an output of zero, then the input is called a "zero of rule F." The inputs where the graph of F touches or crosses the x-axis are the zeros of F [page 40]. When the zeros are not whole numbers, then we can estimate their values [page 41]. This can give us approximate solutions to difficult equations [page 42], and can give us useful information in realistic situations [page 43].

The highest point on the graph of a rule gives us the maximum output of the rule, and the lowest point gives the minimum output. This can be very useful in realistic situations [pages 44-48]. Word problems often ask for the maximum of some quantity. Calculus covers a method for finding exact solutions to such problems, but the basic idea is simply to graph the rule and find the highest point on the graph [page 45]. It is this basic process that we are concerned with in this module. We aren't concerned with finding the exact maximum—a rough estimate that shows that we know the basic process is sufficient. The steps to follow in solving such word problems are outlined on page 45.

If you have trouble with any of the exercises, and if the answers don't give enough help, use the following as a guide for review.

exercises	review
1-3	example 2, page 40
4-6	example 3, page 42
7-9	example 7, page 48
10-11	example 8, page 49
12-15	example 5, page 44, example 6, page 46
12	Module X, pages 20-21 and 38-39
13	Module X, pages 24-26 and page 44
14	Module X, page 46, exercise 19 (plus answer)

Lesson 4

UNDEFINED OUTPUTS AND ASYMPTOTES

EXAMPLE 1 Graph $F(x) = \sqrt{x}$. This will be interesting since there are some inputs for which F has no output.

We pick some easy inputs (ones that have whole-number square roots), make a table of (input,output) pairs, and draw some axes that are long enough to plot all of these pairs. Then we plot them and draw a curve through them.

$F(x) = \sqrt{x}$

input	output
0	0
1	1
4	2
9	3
16	4

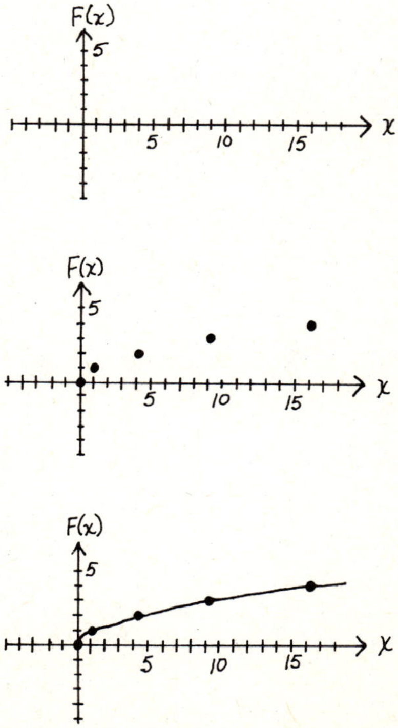

What about negative inputs? The square root of a negative number is a complex number (see Module V, lesson 5). There is no way to show a complex number output on an ordinary graph, so we are restricting outputs to being real numbers. So there are no points to plot where the input is negative. The curve simply ends at 0.

For this rule we also say that if you input a negative number, the output is undefined. But remember that this is just another way of saying that there is no output. The following are different ways of saying the same thing:

There is no output for rule F given input -4.

F(-4) is undefined.

As long as there isn't any output, we can't plot any (input,output) pairs.

Notice that in making our table of inputs and outputs, we only used easy inputs. Why didn't we use 2 or 3 as inputs?

We didn't use them as inputs because the outputs would have been difficult to plot. $F(3) = \sqrt{3}$. How would you plot this output? $\sqrt{3}$ is an irrational number, which means that it is not a whole number or even a fraction. To plot it, you would have to figure out a decimal close to $\sqrt{3}$ that you could plot. This isn't as easy as plotting whole numbers.

But once we have drawn the curve, we can look at it and estimate a decimal close to $\sqrt{3}$ very easily. Can you do this on the graph on the next page?

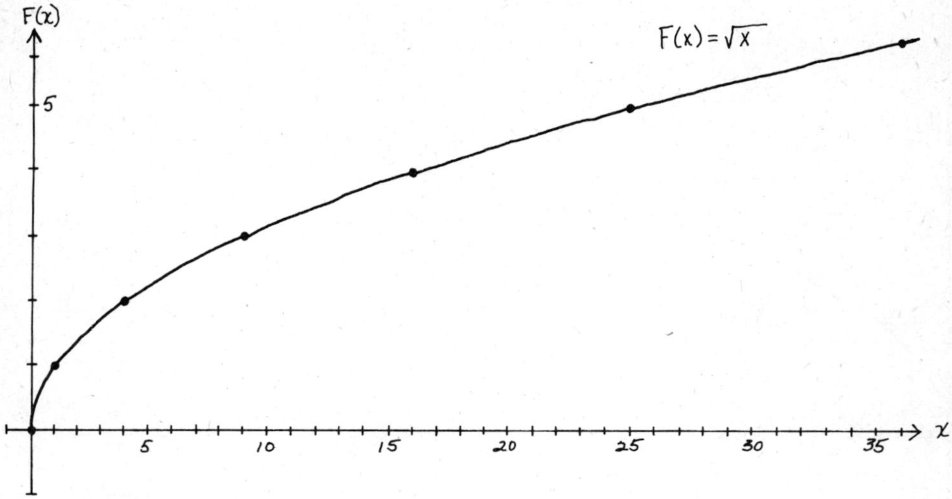

Follow the illustrations below using a straight edge on the graph above.

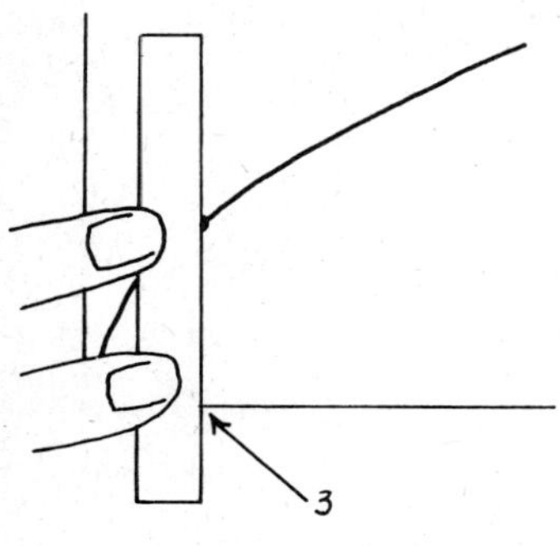

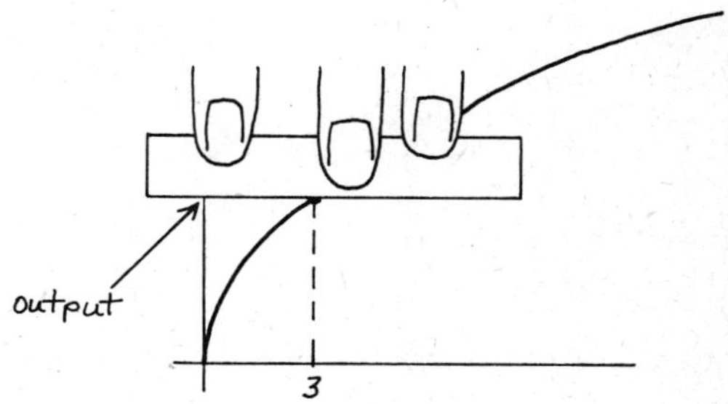

$\sqrt{3}$ is the output of rule F given input 3. So we locate 3 on the x-axis and draw a line straight up until it meets the curve. Mark this point. Now turn the straight edge and draw a line straight over until it meets the F(x)-axis. This marks the output. Can you estimate a decimal close to this?

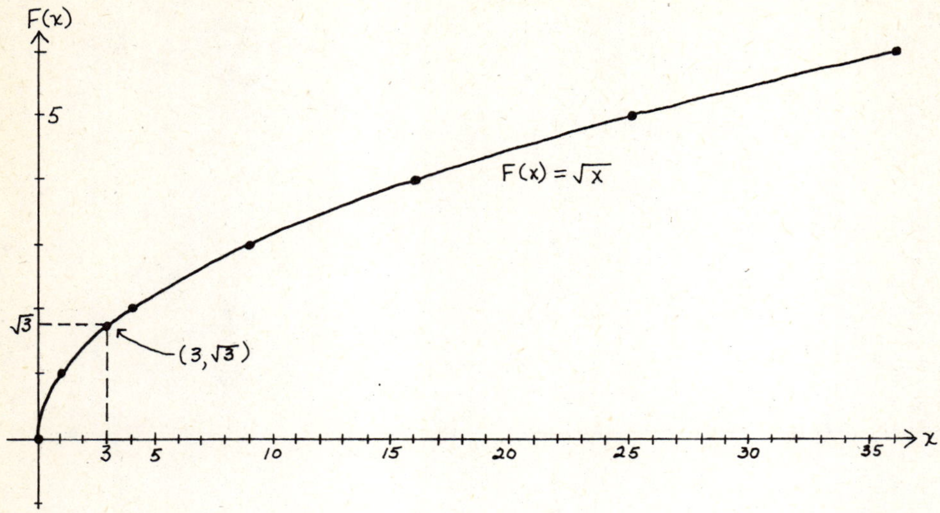

It looks as though the output $\sqrt{3}$ is very close to 1.7. Can we check to see how close?

We can check by arithmetic. We can square 1.7 and see how close it is to 3.

$$(1.7)^2 = 2.89 \quad \text{Not bad.}$$

In chemistry or physics or many of the social sciences, reading graphs accurately is very important. But in mathematics, as we have said, it's usually not very important at all. The most important part of this example is not getting accurate square roots from a graph, but simply knowing that you can do it. <u>You can use a graph to figure outputs</u>. The graph won't give an exact answer, since the accuracy of our eyes and hands is limited. But it's often easier than using a formula.

<u>EXAMPLE 2</u> Here are four more examples of graphs that "end" where there are no more outputs. Work them out for yourself first. Don't spend a long time trying to draw an accurate graph. A quick sketch that shows the right shape, and the point where the curve ends, is all that is needed. Choose easy inputs!

Graph $F(x) = \sqrt{x + 4}$, $G(x) = \sqrt{x - 4}$, $H(x) = \sqrt{4 - x}$, and $J(x) = \sqrt{4 - x} + 3$.

<u>Answers</u>:

$F(x) = \sqrt{x + 4}$

input	output
-4	0
-3	1
0	2
5	3
12	4

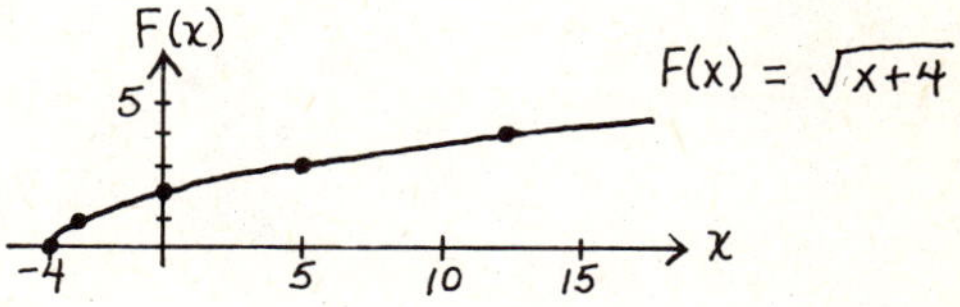

$G(x) = \sqrt{x - 4}$

input	output
4	0
5	1
8	2
13	3
20	4

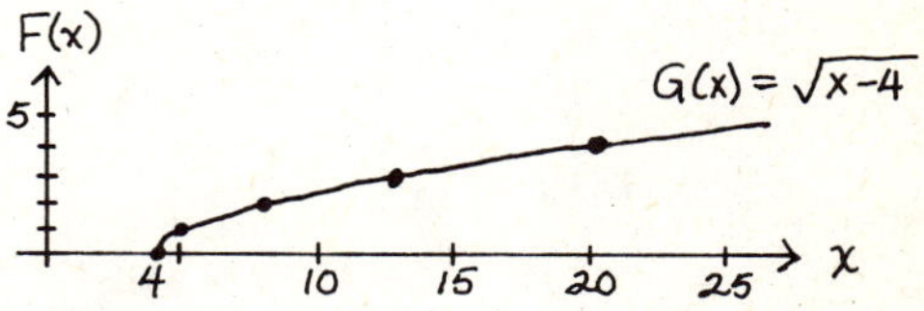

Notice that these two graphs stop on the left but continue on forever on the right.

$H(x) = \sqrt{4 - x}$

input	output
4	0
3	1
0	2
-5	3
-12	4

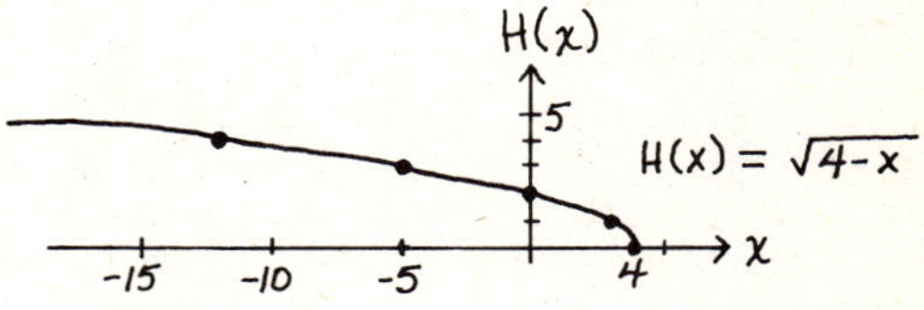

$J(x) = \sqrt{4 - x} + 3$

input	output
4	3
3	4
0	5
-5	6
-12	7

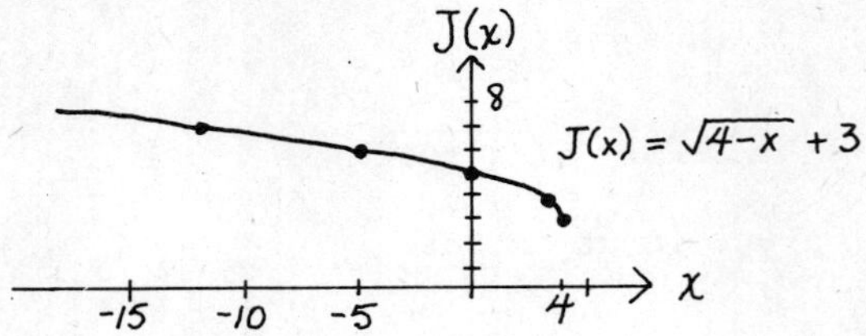

Notice that these two graphs stop on the right but continue on forever on the left.

When the output of a rule is undefined, there is no (input,output) pair to plot on its graph.

EXAMPLE 3 Graph $F(x) = \frac{1}{x^2}$.

Are there any inputs for which the output is undefined? Answer: 0 is the only one. F(0) is undefined. Does this mean that the graph of F ends at 0? Well, in a way it does, but as we shall soon see, it looks somewhat different from the graph of a square root rule.

$F(x) = \frac{1}{x^2}$

input	output
-3	$\frac{1}{9} \approx 0.1$
-2	$\frac{1}{4} = 0.25$
-1	1
1	1
2	$\frac{1}{4} = 0.25$
3	$\frac{1}{9} \approx 0.1$

These points are plotted on the axes below. Make a guess as to what the graph looks like by drawing it in with your pencil. Then check your curve before you go on.

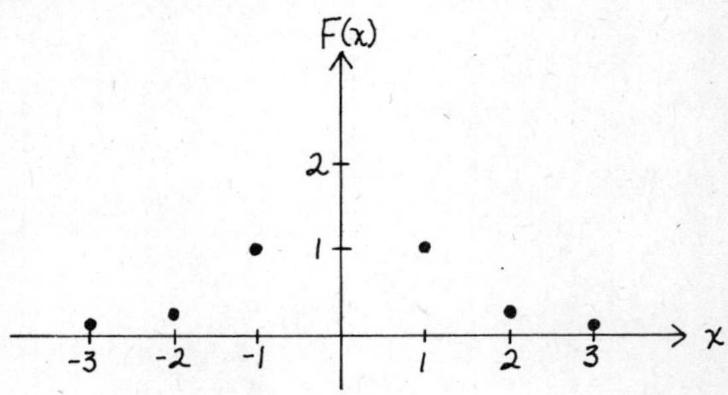

You might have been tempted to draw the curve something like this:

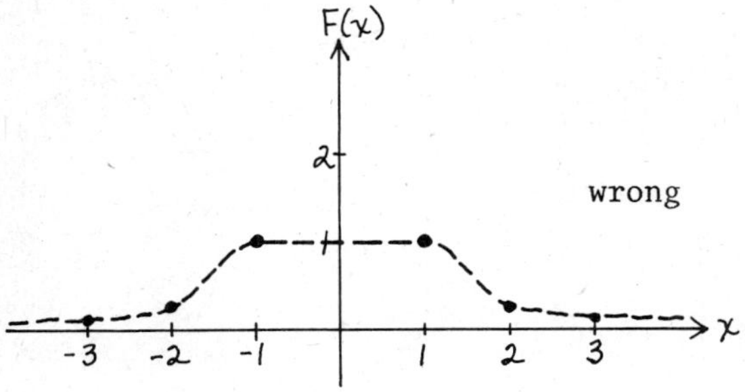

If you were, you were giving the rule an output of 1 for input 0. But we know that there is <u>no</u> (input,output) pair when the input is 0. So this curve can't be correct.

Let's check the curve by plotting more points near the troublesome input 0.

$$F(x) = \frac{1}{x^2}$$

input	output	
$-\frac{1}{2}$	4	Work this out!
$-\frac{1}{3}$	9	Work this out!
$\frac{1}{2}$	4	
$\frac{1}{3}$	9	

We'll need larger numbers on the F(x)-axis to plot these, so let's re-draw our graph. These points are plotted on the next page. Can you draw the curve through these points?

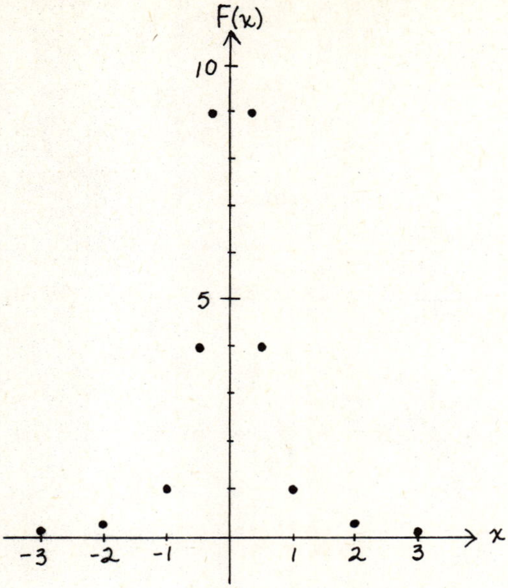

You might be tempted to draw the graph below:

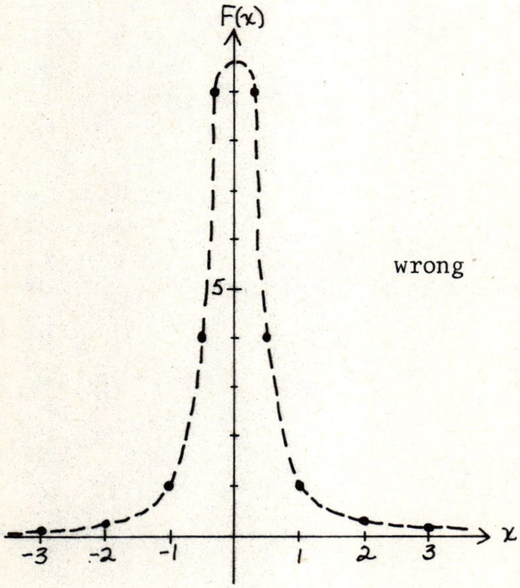

But remember, there is no output for input 0. The only way to keep the graph from showing an output for input 0 is to keep the graph from crossing the F(x)-axis!

This means that the curve must be in two pieces that never meet!

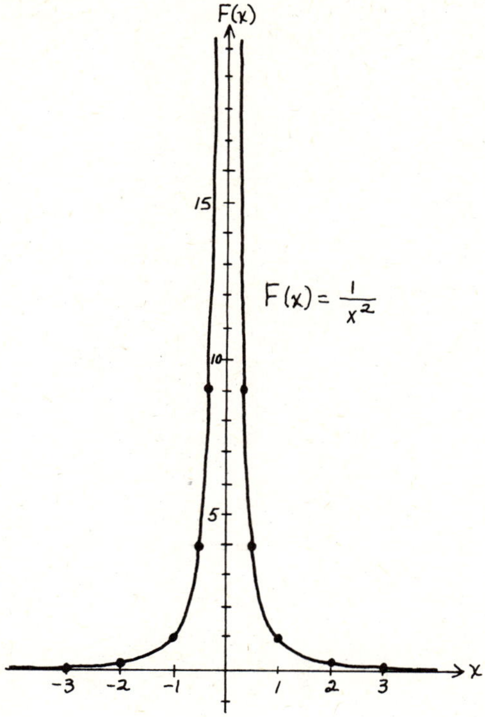

Check this curve by plotting points for inputs $\frac{1}{4}$ and $-\frac{1}{4}$.

<u>This graph never touches the F(x)-axis.</u> If it did, it would indicate an output for input 0, and that can't happen.

The graph below is also a graph of $F(x) = \frac{1}{x^2}$. The F(x)-axis (which we have left out of the picture) goes up to 10,000. The left and right sides of the graph do not meet.

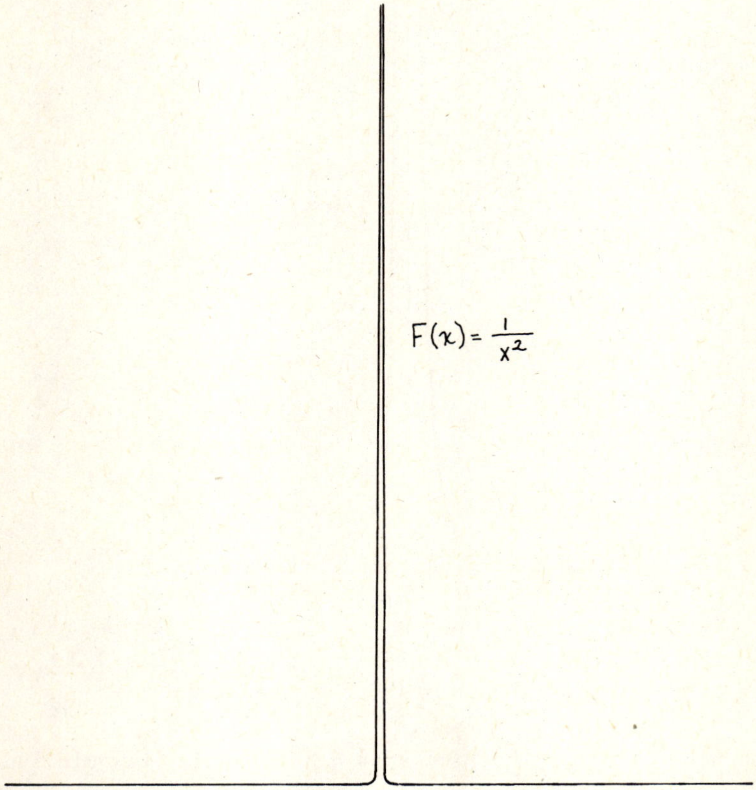

An algebraic way to see why the graph looks like this is to think of using smaller and smaller fractional inputs to $F(x) = \frac{1}{x^2}$ (such as $\frac{1}{2}$, $\frac{1}{4}$, $\frac{1}{8}$, etc.) You can see that the outputs get larger and larger without limit.

Seg. 9 - skip to p. 82

ASYMPTOTES

If we trace along the left-hand side of the graph of $F(x) = \frac{1}{x^2}$, we see that it gets closer and closer to the F(x)-axis, but it never reaches this axis.

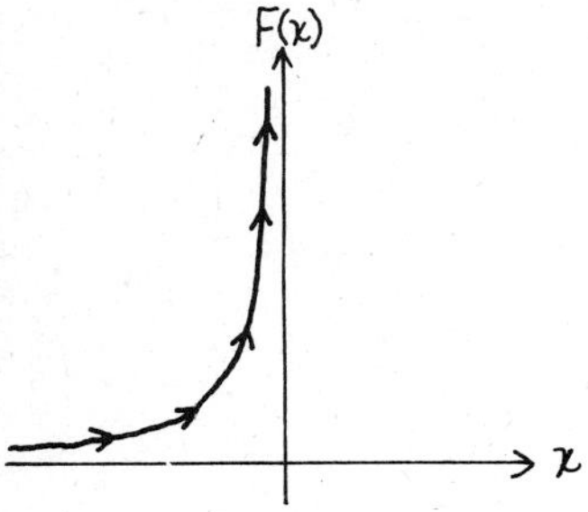

(The same is true for the right-hand side.) There is a special word for this situation. In this case, we say that the F(x)-axis is an asymptote to the graph.

> Whenever a graph gets closer and closer to a straight line, without ever reaching the line (as in the picture above), then the line is called an asymptote to the graph.

If we trace either side of the graph of $F(x) = \frac{1}{x^2}$ away from the F(x)-axis, we see that the curve gets closer and closer to the x-axis without ever reaching it. For instance, for the right-hand side:

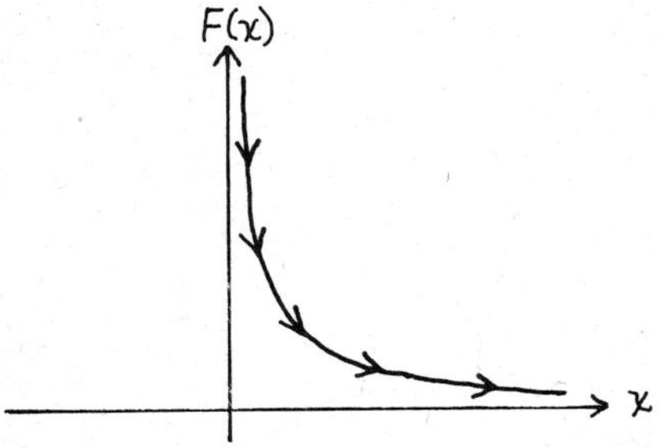

So the x-axis is also an asymptote to this graph. How can we be sure that the graph will never reach the x-axis? If it did, then at the point where it touched the x-axis , the output would be zero. But $\frac{1}{x^2}$ does not have any zeros (as we can see if we try to solve $\frac{1}{x^2} = 0$ algebraically).

Since we often think of the F(x)-axis as being vertical, we say that the F(x)-axis is a vertical asymptote to the graph of $F(x) = \frac{1}{x^2}$. We say that the x-axis is a horizontal asymptote to this graph.

EXAMPLE 4 Graph $F(x) = \frac{1}{x}$.

$$F(x) = \frac{1}{x}$$

input	output
-10	$-\frac{1}{10}$
-4	$-\frac{1}{4}$
-2	$-\frac{1}{2}$
-1	-1
1	1
2	$\frac{1}{2}$
4	$\frac{1}{4}$
10	$\frac{1}{10}$

When you plot these points, you might be tempted to draw the curve below. But just as in the last example, the curve can't cross the F(x)-axis, since there is no output given input 0.

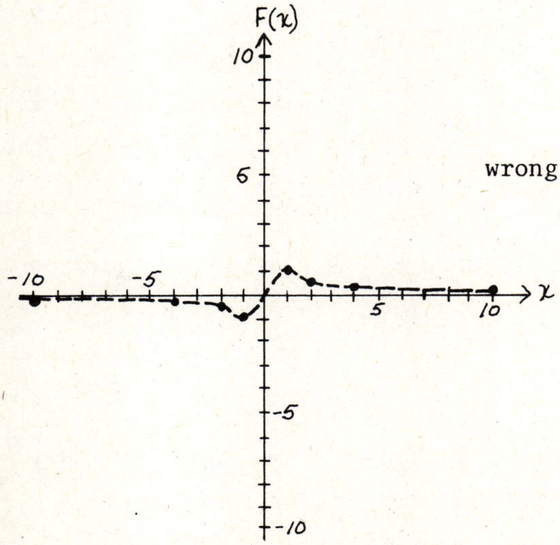

wrong

So we should plot some more points using inputs between -1 and 1.

$$F(x) = \frac{1}{x}$$

input	output
$-\frac{1}{2}$	-2
$-\frac{1}{4}$	-4
$-\frac{1}{10}$	-10
$\frac{1}{10}$	10
$\frac{1}{4}$	4
$\frac{1}{2}$	2

When we plot these points, in addition to the ones we already have, the shape of the curve becomes more apparent.

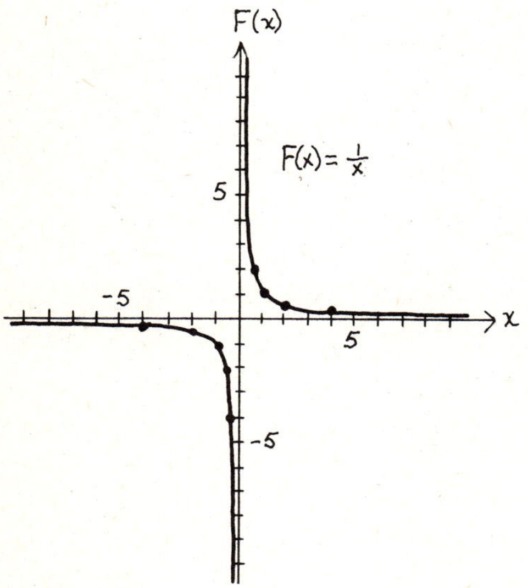

This time one side of the curve goes up while the other goes down. Neither piece of the graph ever touches the F(x)-axis.

Again, the F(x)-axis is a vertical asymptote, and the x-axis is a horizontal asymptote.

If we had known in advance that the F(x)-axis was going to be an asymptote, then the graph would have been easier to draw. We wouldn't have had to plot as many points to see the shape of the graph. For instance:

From these points alone (without the fractional inputs),

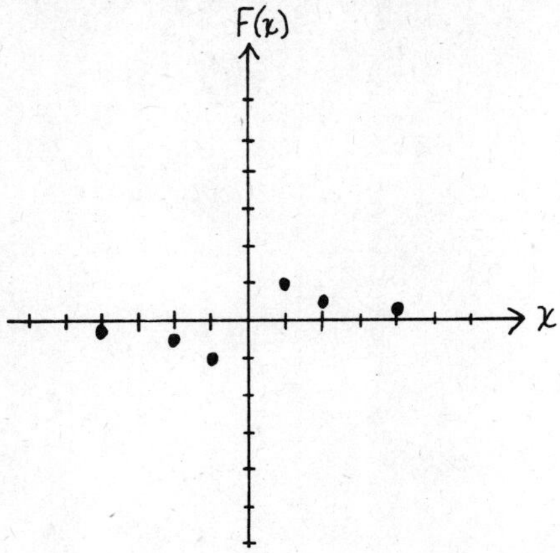

we can be fairly certain of this much of the curve:

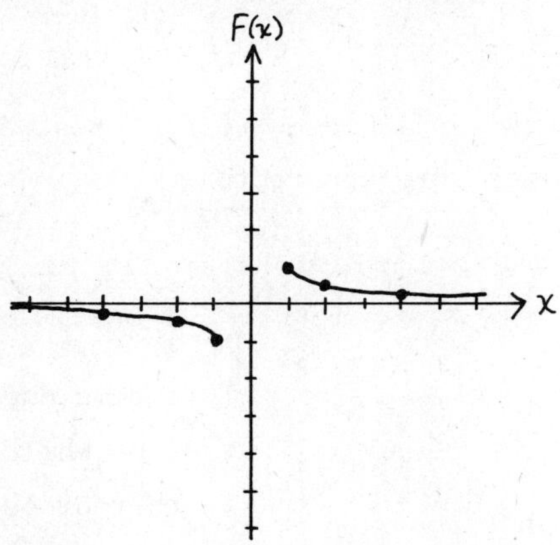

and if we know that the F(x)-axis is an asymptote, then we can be fairly certain that the curve continues as follows:

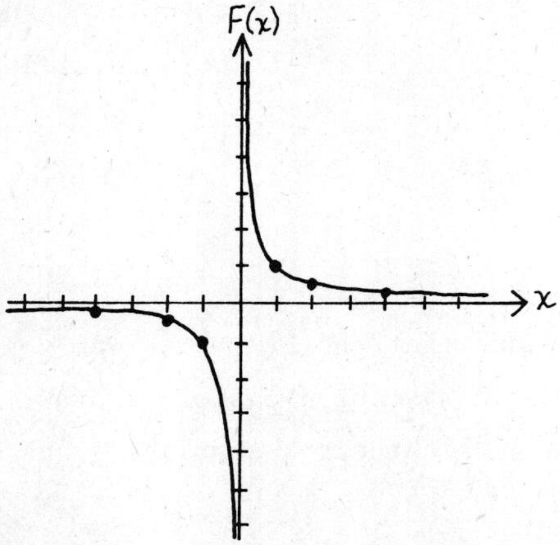

How can we tell in advance whether or not a graph will have a vertical asymptote? In this case, we could tell by looking at the denominator of $\frac{1}{x}$. When the denominator is zero, the graph has a vertical asymptote.

> Usually (but not always), a graph has a vertical asymptote whenever the denominator in the formula is zero.

<u>EXAMPLE 5</u> Graph $F(x) = \frac{1}{x - 4}$.

Will the graph of F have a vertical asymptote? The denominator is x - 4, and this is zero when x = 4. So the graph of F may have a vertical asymptote through x = 4:

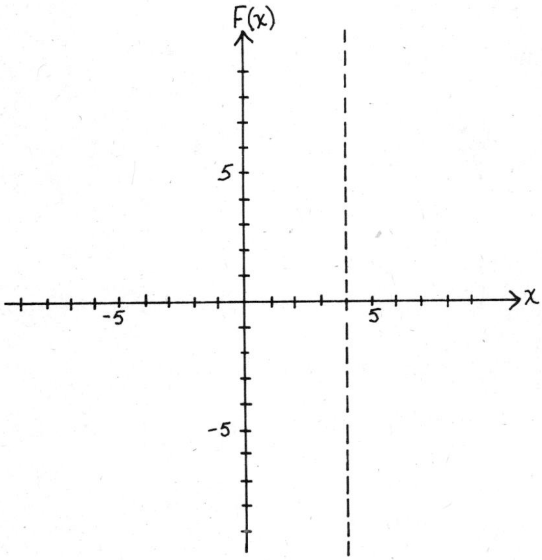

We know that the graph will not touch the dotted line, since F has no output when the input is 4. Try drawing the graph of F before you read on. You can check whether or not the dotted line through x = 4 is an asymptote by trying some inputs close to 4.

Answer:

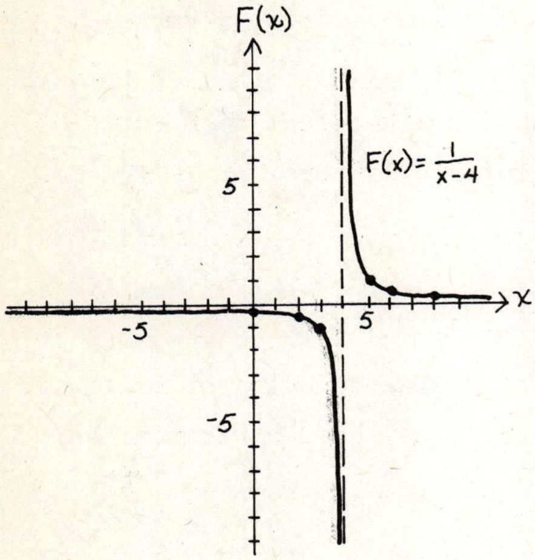

Again, this graph has two asymptotes: the vertical line through x = 4, and the horizontal x-axis. Knowing about the vertical asymptote in advance made the graph easier to draw.

EXAMPLE 6 Graph $G(x) = \frac{1}{x + 4}$. Draw the vertical asymptote first.

Answer:

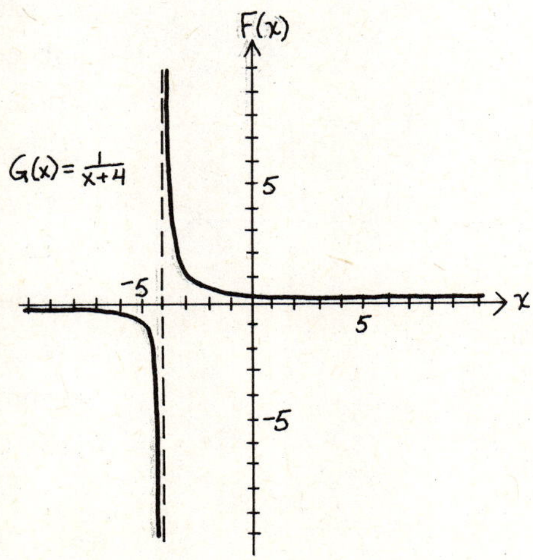

EXAMPLE 7 Graph $F(x) = x + \frac{1}{x}$.

Notice that one part of the formula has a denominator that is zero when x = 0. This suggests that the vertical line through x = 0 will be a vertical asymptote—that is, the F(x)-axis will be a vertical asymptote.

$$F(x) = x + \frac{1}{x}$$

input	output
-4	$-4\frac{1}{4}$
-3	$-3\frac{1}{3}$
-2	$-2\frac{1}{2}$
-1	-2
0	undefined
1	2
2	$2\frac{1}{2}$
3	$3\frac{1}{3}$
4	$4\frac{1}{4}$

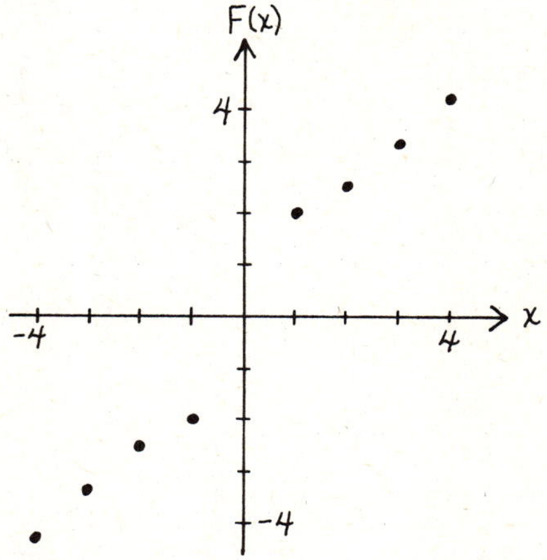

But now which way does the curve bend?

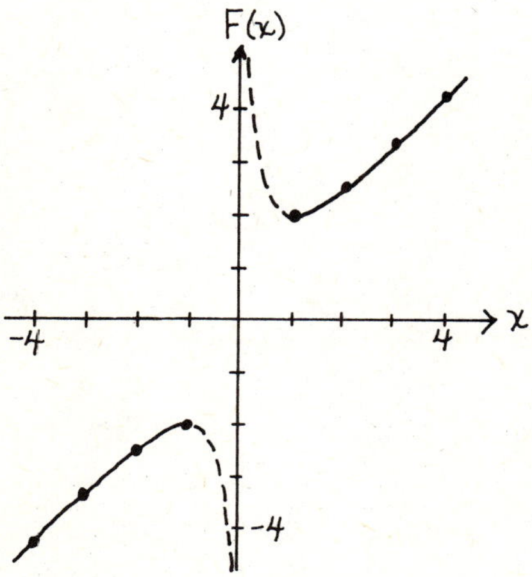

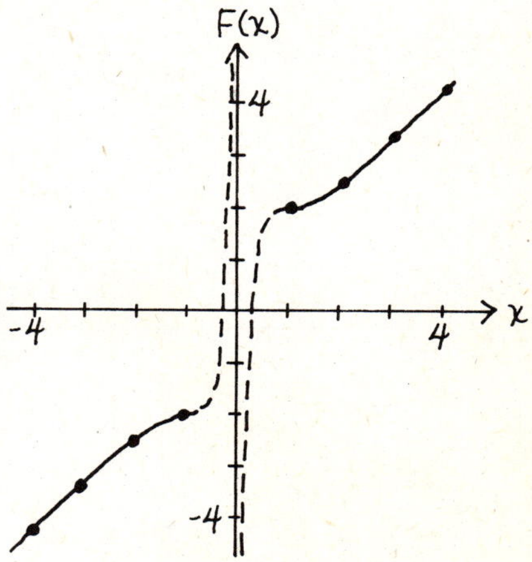

The first way looks the most likely, but we should check with some fractional inputs to make sure.

$F(x) = x + \frac{1}{x}$

input	output
$-\frac{1}{2}$	$-2\frac{1}{2}$
$-\frac{1}{4}$	$-4\frac{1}{4}$
$\frac{1}{4}$	$4\frac{1}{4}$
$\frac{1}{2}$	$2\frac{1}{2}$

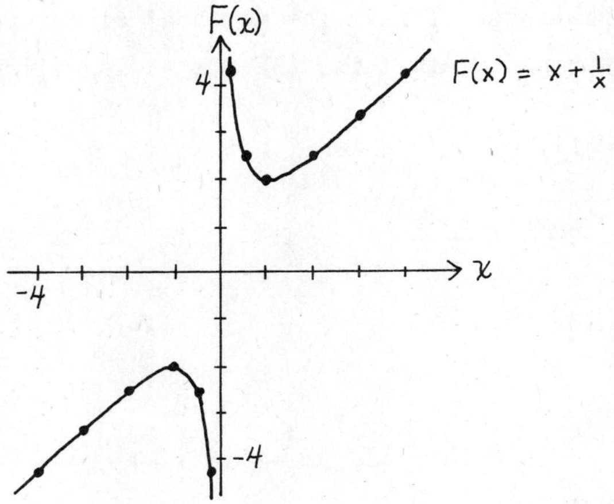

EXAMPLE 8 Graph $F(x) = \frac{6}{x^2 - 4}$.

Is the denominator ever zero? To find out, we solve $x^2 - 4 = 0$. This factors into $(x + 2)(x - 2) = 0$, so the denominator is zero when $x = -2$ and when $x = 2$. There will be two vertical asymptotes!

$F(x) = \frac{6}{x^2 - 4}$

input	output
-4	0.5
-3	1.2
-2	undef.
-1	-2
0	-1.5
1	-2
2	undef.
3	1.2
4	0.5

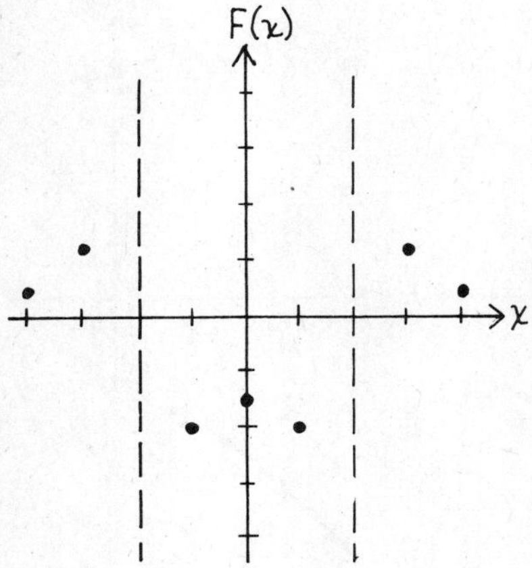

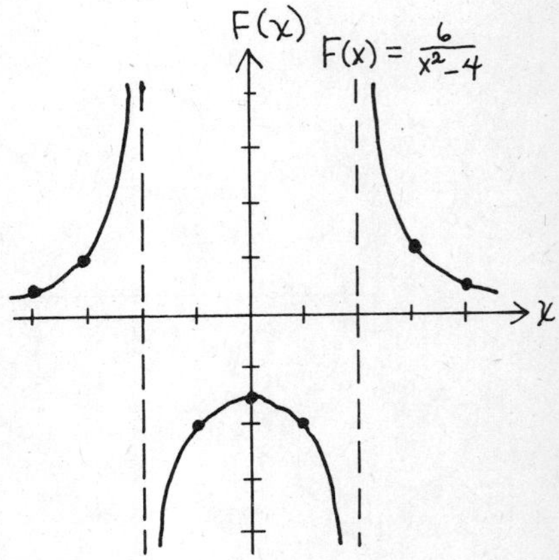

Without knowing the vertical asymptotes in advance, we would have had to plot many more points in order to draw this graph.

EXERCISES (Answers start on page 78.)

Sketch graphs of the following rules.

1. $F(x) = \sqrt{x}, \quad G(x) = -\sqrt{x}$
2. $F(t) = \sqrt{3 + t}, \quad G(t) = \sqrt{3 - t}$
3. $F(s) = \sqrt{s - 3}, \quad G(s) = 2 + \sqrt{s - 3}$
4. $F(u) = \dfrac{1}{u + 2}, \quad G(u) = \dfrac{-1}{u + 2}$
5. $F(v) = \dfrac{1}{v - 2}, \quad G(v) = \dfrac{1}{2 - v}$
6. $F(w) = \dfrac{1}{(w - 1)^2}, \quad G(w) = \dfrac{1}{(w - 1)^2} + 1$
7. $F(z) = \dfrac{1}{z^2 - 1}, \quad G(z) = \dfrac{1}{1 - z^2}$
8. $F(r) = \dfrac{1}{r} - r, \quad G(r) = r - \dfrac{1}{r}$
9. $F(s) = \dfrac{2}{s} + \dfrac{s}{2}, \quad G(s) = \dfrac{2}{s} + \dfrac{s}{2} +$ [redacted] 1
10. $F(t) = \sqrt{25 - t^2}, \quad G(t) = -\sqrt{25 - t^2}$
11. $F(x) = \dfrac{1}{\sqrt{4 + x}}, \quad G(x) = \dfrac{1}{\sqrt{4 - x}}$

ANSWERS TO EXERCISES

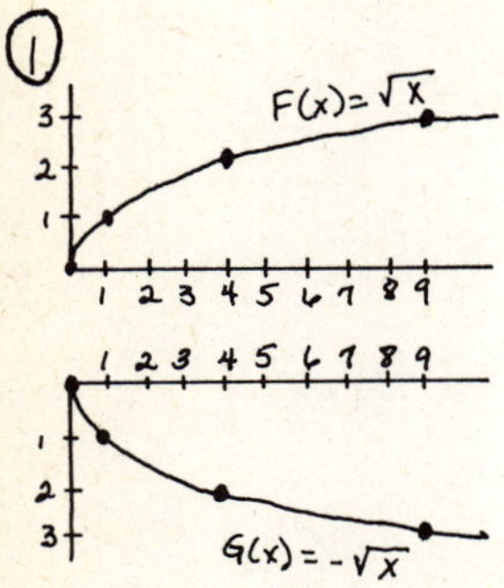

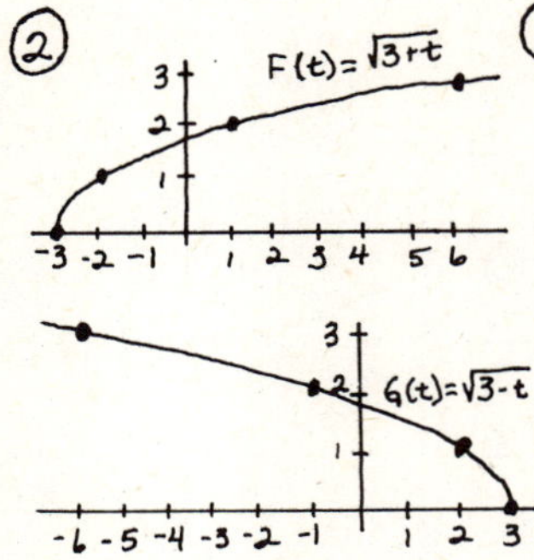

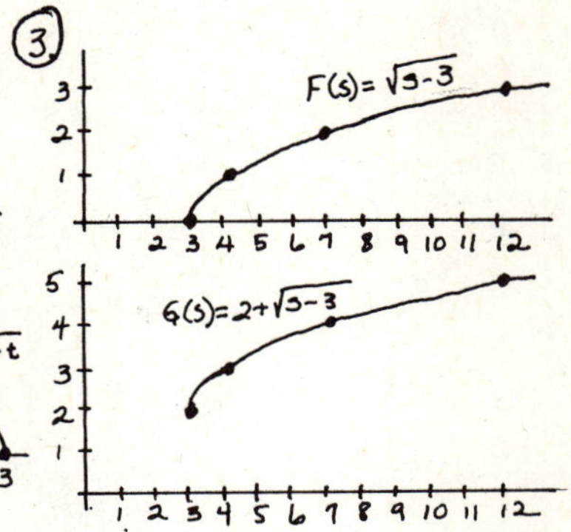

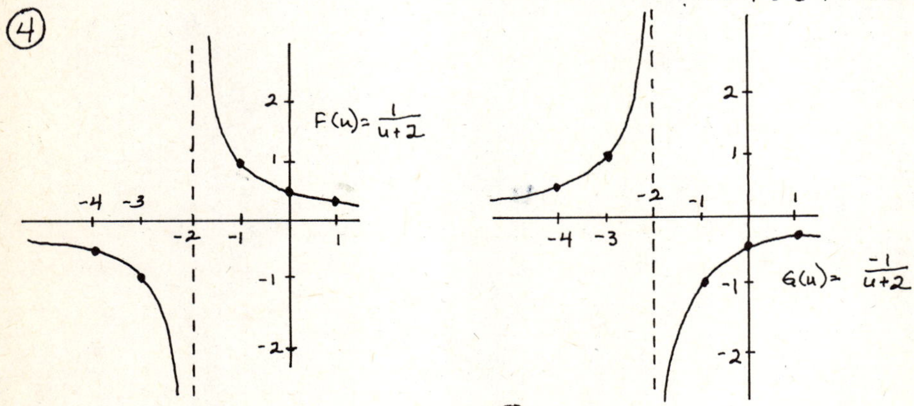

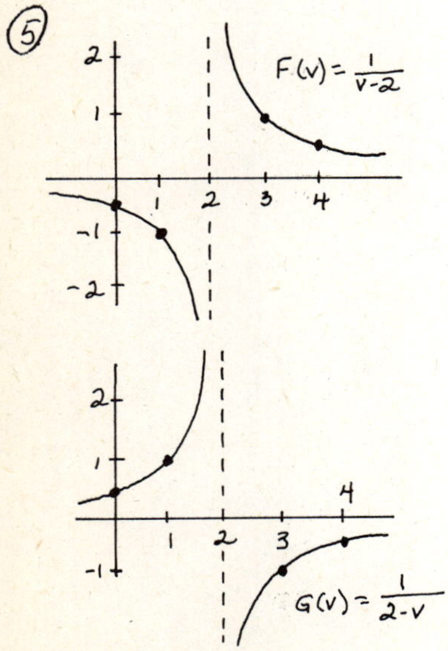

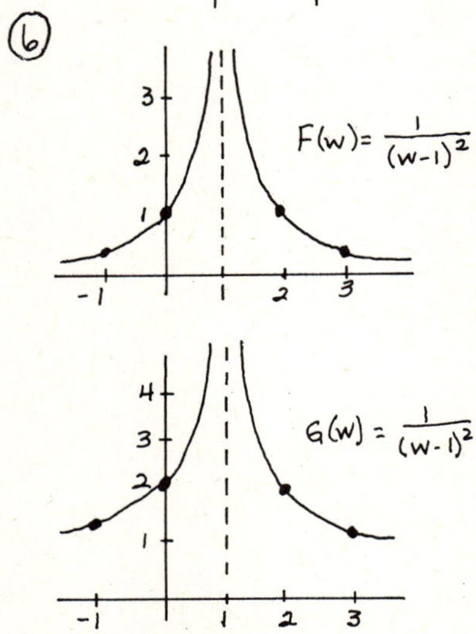

(7)

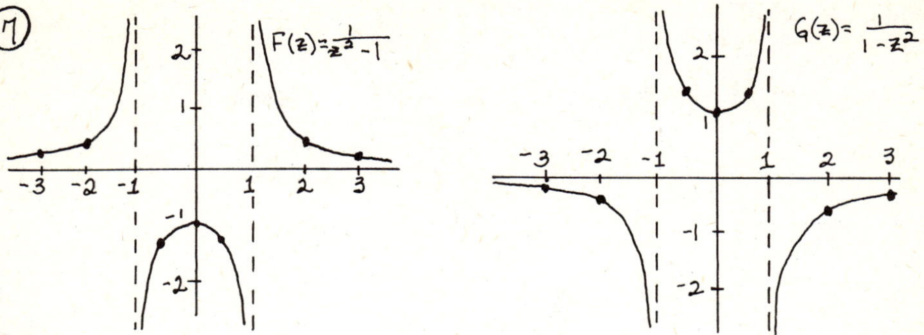
F(z) = 1/(z² - 1)
G(z) = 1/(1 - z²)

(8)

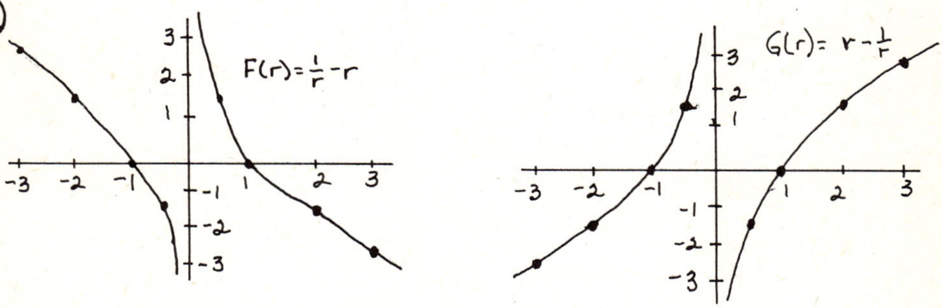
F(r) = 1/r - r
G(r) = r - 1/r

(9)

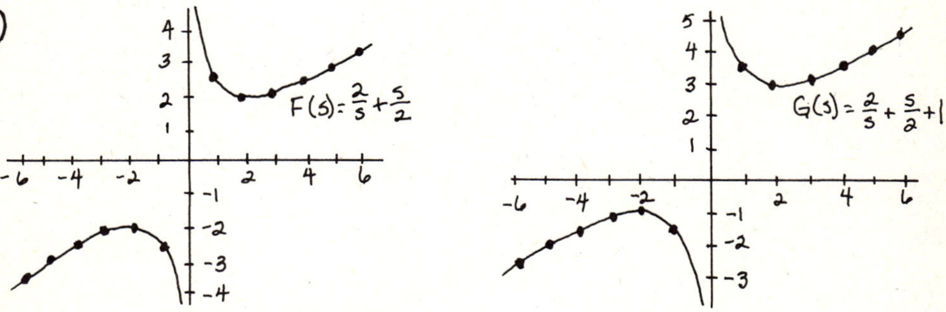
F(s) = 2/s + s/2
G(s) = 2/s + s/2 + 1

(10)

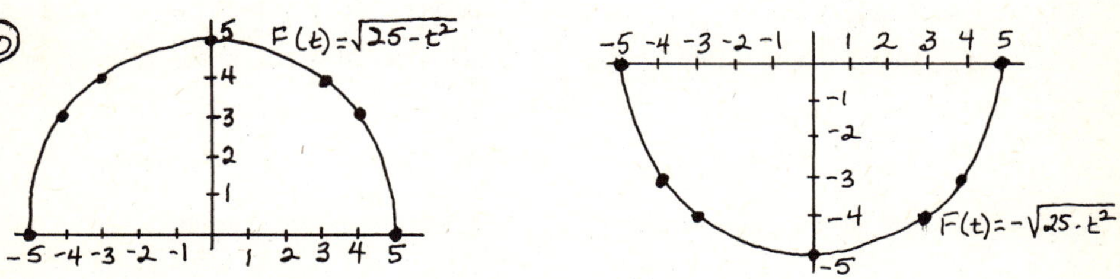
F(t) = √(25 - t²)
F(t) = -√(25 - t²)

(11)

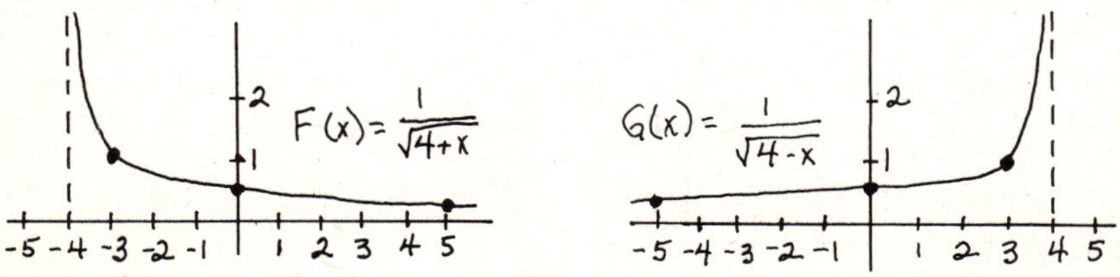
F(x) = 1/√(4+x)
G(x) = 1/√(4-x)

ADDITIONAL EXERCISES

Sketch the graphs of the following rules.

1. $F(x) = \sqrt{x + 1}$, $G(x) = -\sqrt{x + 1}$
2. $F(t) = \sqrt{2 + t}$, $G(t) = \sqrt{2 - t}$
3. $F(s) = \sqrt{s - 2}$, $G(s) = 3 + \sqrt{s - 2}$
4. $F(u) = \frac{1}{u + 3}$, $G(u) = \frac{1}{u + 3}$
5. $F(v) = \frac{1}{v - 3}$, $G(v) = \frac{1}{3 - v}$
6. $F(w) = \frac{1}{(w + 1)^2}$, $G(w) = \frac{1}{(w + 1)^2} + 1$
7. $F(z) = \frac{9}{z^2 - 9}$, $G(z) = \frac{9}{9 - z^2}$
8. $F(r) = \frac{1}{(r + 1)} - (r + 1)$, $G(r) = (r + 1) - \frac{1}{(r + 1)}$
9. $F(s) = \frac{3}{s} + \frac{s}{3}$, $G(s) = \frac{3}{s} + \frac{s}{3} + 1$
10. $F(t) = \sqrt{16 - t^2}$, $G(t) = -\sqrt{16 - t^2}$
11. $F(x) = \frac{1}{\sqrt{1 + x}}$, $G(x) = \frac{1}{\sqrt{1 - x}}$

SUMMARY OF LESSON 4

This lesson introduces the graphs of square root functions, such as $F(x) = \sqrt{x}$, and reciprocal functions, such as $F(x) = \frac{1}{x}$. When the output of a rule is undefined, there is no (input,output) pair to plot on its graph. As a result, the graphs of square root functions simply end

 where the expression under the square root sign becomes negative. The graphs of reciprocal functions "blow up," 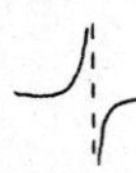 or 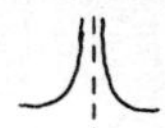, where the denominator is zero, separating the graph into disconnected pieces.

"Vertical and horizontal asymptotes" are defined on page 69, and the rest of the examples illustrate how to predict vertical asymptotes and how we use them to help draw the graph.

If you have trouble with any of the exercises, and if the answers don't provide enough help, use the following as a guide for review.

exercises	review
1-3	example 1, page 59; example 2, page 62
4-5	examples 4-6, pages 70-74
6	example 3, page 64
7	example 8, page 76
8-9	example 7, page 75

Lesson 5

THE DOMAIN OF A RULE

FINDING DOMAINS ALGEBRAICALLY

EXAMPLE 1 rule F: $F(x) = \frac{1}{x}$

Which inputs to rule F give outputs, and which do not?

We know that 0 is the only "bad" input to this rule. When the input is 0, the output is undefined. Or to say it another way, when the input is 0, there is no output. But every real number except 0 is a "good" input. If we use any real number other than 0 as an input, we can follow rule F and get an output.

F(0) is undefined, but $F(2) = \frac{1}{2}$, $F\left(\frac{1}{3}\right) = 3$, $F(-1) = -1$, $F(\sqrt{2}) = \frac{1}{\sqrt{2}}$, and so on.

There is some special mathematical vocabulary for this situation. When we want to talk about the "good" inputs to a rule, we use the word domain.

> The domain of an input-output rule is all inputs to the rule that give outputs.

So, the domain of rule F above is all real numbers except 0.

EXAMPLE 2 rule G: $G(t) = \frac{1}{t^2 - 4}$

What is the domain of this rule?

We can divide by any number except 0, so the domain of rule G will be all real numbers except the ones that make the denominator come out 0. To find out which numbers make the denominator zero, we must solve the equation $t^2 - 4 = 0$.

$$t^2 - 4 = 0$$
$$(t + 2)(t - 2) = 0$$

The solutions are 2 and -2, so 2 and -2 make the denominator of the formula for G come out zero.

Answer: The domain of rule G is all real numbers except -2 and 2.

EXAMPLE 3 rule H: $H(s) = \sqrt{s}$
What is the domain of this rule?

We can take the square root of positive numbers, and 0 (the output is 0). But we can't take the square root of a negative number. (Remember that we are restricting inputs and outputs to being real numbers.)

Answer: The domain of rule H is all real numbers greater than 0, and 0, or all nonnegative real numbers or all real numbers greater than or equal to 0.

EXAMPLE 4 rule F: $F(x) = \sqrt{x + 4}$
What is the domain of this rule?

We just have to make sure that we don't get a negative number under the square root sign. This means that x can be -4 (since $\sqrt{0}$ is OK), but x can't be more negative than -4.

Answer: The domain of rule F is all real numbers greater than -4, and -4, or all real numbers greater than or equal to -4.

EXAMPLE 5 rule G: $G(x) = \frac{1}{\sqrt{x + 4}} - 2$

What is the domain of this rule? who cares?

Here, we have to watch out for two things—no negative under the square root, and no zero in the denominator. This means that -4 is now a bad input, since it makes the denominator come out 0.

Answer: The domain of rule G is all real numbers greater than -4.

For now, there are only two things to watch out for when finding the domain of a rule: no negative under a square root sign, and no zero in any denominator. (When you study exponentials, logarithms, and trigonometry, there will be a few more things to watch out for.)

FINDING DOMAINS GRAPHICALLY

We can also find the domain of a rule by looking at its graph.

EXAMPLE 6 rule F: $F(x) = \sqrt{x + 4}$

Find the domain of this rule by looking at its graph. (We found the domain of rule F algebraically in example 4. This rule was graphed on page 63.)

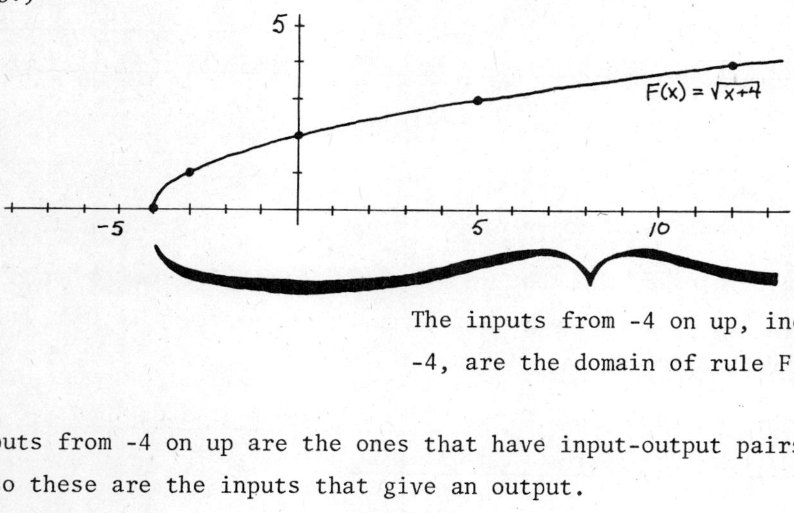

The inputs from -4 on up, including -4, are the domain of rule F.

The inputs from -4 on up are the ones that have input-output pairs above them, so these are the inputs that give an output.

Another way to indicate the domain is to shade the part of the x-axis that lies directly below the graph:

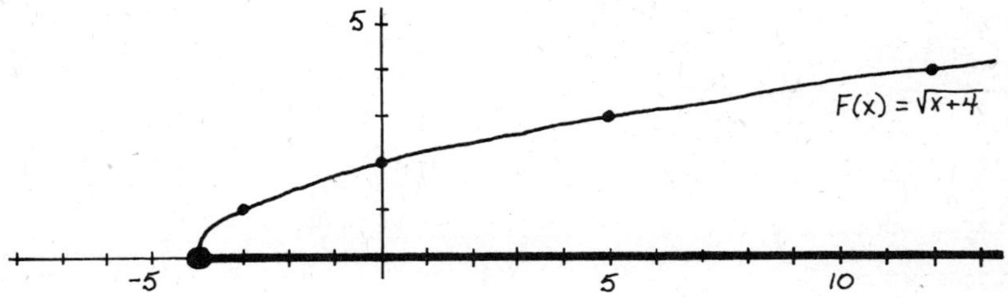

The big solid dot at -4 indicates that -4 is part of the domain. The shading shows that every number greater than -4 is also part of the domain.

EXAMPLE 7 rule G: $G(x) = \frac{1}{\sqrt{x + 4}} - 2$

Sketch the graph of this rule, and indicate its domain by shading the input axis.

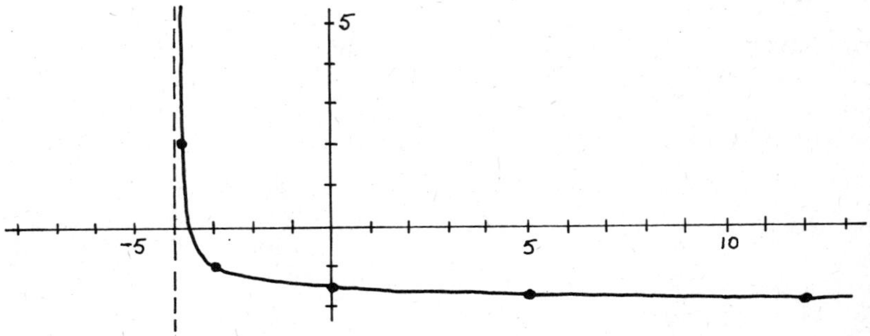

Rule G has no output given input -4 (the graph has a vertical asymptote there). The domain is shaded below:

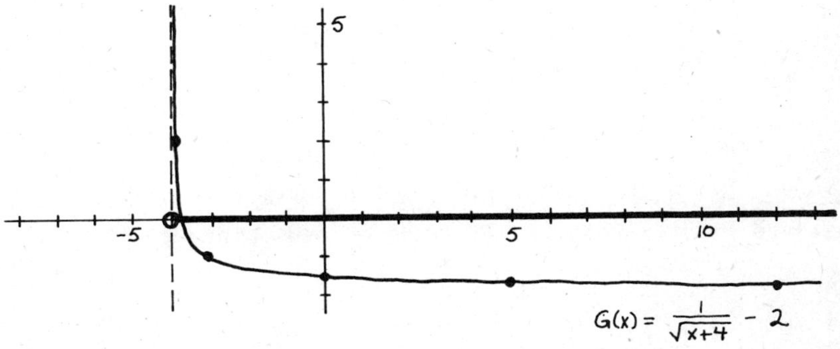

The big hollow dot at -4 indicates that -4 is not part of the domain. The shading indicates that every number greater than -4 is part of the domain. This convention is used in many math books. A solid dot at the end of some shading indicates that the end number is included, while a hollow dot means that the end number is not included.

As the last two examples illustrate, the part of the input axis that corresponds to the domain of a rule is the part that lies directly above or below the graph.

When you want to find the domain of a rule, should you do it algebraically (as in examples 1-5) or graphically (as in examples 6-7)? If you already have the graph of the rule on hand, then it will probably be easiest to use it. Otherwise, it will usually be faster to do it algebraically. If the algebraic method gets confusing, then sketching the graph may help.

INEQUALITY NOTATION

Up to this point, we have been describing the domains of rules in one of two ways. We have been writing out the description in English, such as "all numbers greater than -4," or we have been drawing an x-axis and shading the appropriate values of x. There is a mathematical shorthand that can save a lot of this writing.

EXAMPLE 8 The following all indicate the same numbers:

✓ English: all numbers greater than -4

✓ picture of x-axis:

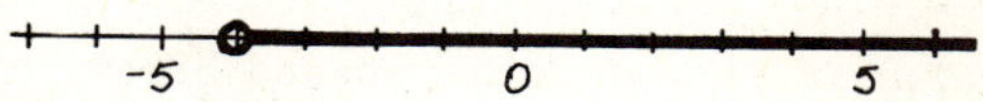

✓ mathematical language: all values of x greater than -4

✓ inequality notation: $x > -4$ (read "x is greater than -4")

The little wedge $>$ is called an "inequality sign." The wide end of it always faces the larger quantity. For instance, $x > -4$ indicates values of x that are greater than -4. The narrow end points to the smaller quantity.

We could also indicate the same numbers by writing $-4 < x$, which would be read "-4 is less than x." Again, the narrow end of the wedge points to the smaller side, and the wide end points to the larger side.

EXAMPLE 9 How could we include -4, along with all numbers greater than -4? (The rule in examples 4 and 6 had this domain.)

✓ English: all numbers greater than -4, and -4

✓ picture of x-axis:

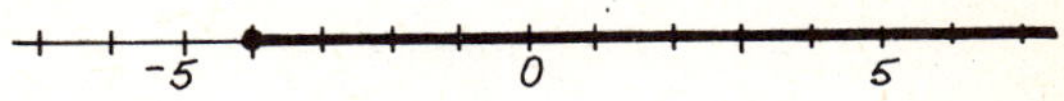

✓ mathematical language: all values of x greater than or equal to -4

✓ inequality notation: $\boxed{x \geq -4}$ Read "x is greater than or equal to -4."

Or, we could have written $\boxed{-4 \leq x}$ to indicate the same numbers. The important thing is that the narrow end of the inequality sign points to the smaller quantity, and the wide end to the larger.

EXAMPLE 10 How can we indicate all numbers between -4 and 3, not including either one?

✓ English: all numbers between -4 and 3, not including -4 or 3

✓ picture of x-axis:

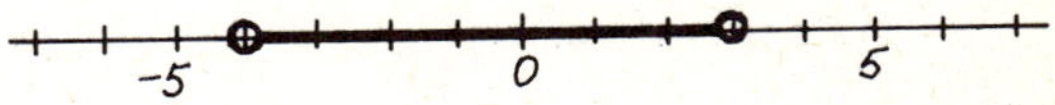

✓ mathematical language: all values of x greater than -4 and less than 3

✓ inequality notation: $\boxed{-4 < x < 3}$ Read "-4 is less than x is less than 3."

Note: When using two inequality signs together like this, we put the -4 on the left and the 3 on the right. It is customary to put the smaller number on the left and the greater number on the right, just as they appear on the x-axis. Then we have to be sure to point the narrow ends of the wedges toward the left, toward the smaller quantity.

EXAMPLE 11 In each case, describe the indicated numbers by using inequality notation.

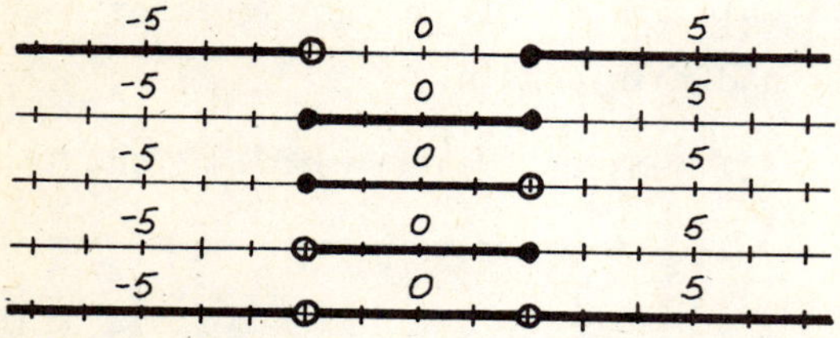

	picture	inequality notation
a.		$x < -2, \quad x \geqslant 2$
b.		$-2 \leqslant x \leqslant 2$
c.		$-2 \leqslant x < 2$
d.		$-2 < x \leqslant 2$
e.		$x < -2, \ -2 < x < 2, \ x > 2$

Notes: 1. As mentioned above, when using two inequality signs in the same expression, it is customary to put the smaller number on the left and the larger number on the right (just as they appear on the x-axis). Consequently, we have to point the narrow end of the wedges to the left, toward the smaller side.

2. When using only one inequality sign in an expression, it is customary to put the x on the left, and then decide which way the wedge should point. In part a, we could have written $-2 > x$, $2 \leqslant x$. But putting the x on the left is conventional.

3. In part e, the picture indicates all numbers except -2 and 2. Another mathematical shorthand for this, besides the one given, is $\boxed{x \neq -2, 2}$, read "x does not equal -2 or 2."

Study the example above carefully, until you are sure you understand how to decide which direction to point the wedges, and how to decide whether to use $\leqslant$ or $<$. Make sure you can write out the inequalities yourself without looking at the answers in this module.

EXAMPLE 12 Indicate the domains of the following rules by using inequality notation.

rule	domain
$F(x) = \frac{1}{x}$	$x < 0,\ x > 0$ (or $x \neq 0$)
$F(x) = \sqrt{x + 4}$	$x \geqslant -4$
$G(x) = \frac{1}{\sqrt{x + 4}} - 2$	$x > -4$

We found these domains before in examples 1, 4, and 5.

All the formulas so far used the letter x to stand for the input, and we used the letter x in the inequalities that described the domain. If a formula uses some other letter to stand for the input, then we must use that letter in our inequalities.

EXAMPLE 13 Indicate the domain of the following rules by using inequality notation.

rule	domain
$F(u) = \frac{1}{u}$	$u \neq 0$ (or $u < 0,\ u > 0$)
$H(s) = \sqrt{s}$	$s \geqslant 0$
$G(t) = \frac{1}{t^2 - 4}$	$t \neq -2, 2$ (or $t < -2,\ -2 < t < 2,\ t > 2$)

We found these domains before in examples 1, 3, and 2. A picture of the domain of rule G appears in example 11, part e.

EXAMPLE 14 Indicate the domain of the following rules by using inequality notation.

rule	domain
$F(x) = x + 1$	all real numbers
$G(x) = \frac{(x + 1)(x - 1)}{(x - 1)}$	$x \neq 1$

Notice that if we try to compute G(1), we get $\frac{2(0)}{0}$. This is undefined. We can't cancel the 0's. 1 is not in the domain of rule G.

It is tempting to cancel the $(x - 1)$'s in the formula for rule G. But then we would get $G(x) = x + 1$, and this formula would give us the wrong answer for $G(1)$. $G(1)$ is undefined, as we saw above.

So it is not quite correct to write $\frac{(x + 1)(x - 1)}{(x - 1)} = x + 1$. Technically, we should write: $\frac{(x + 1)(x - 1)}{(x - 1)} = x + 1$ <u>as long as x does not equal 1</u>.

EXERCISES (Answers start on page 92.)

Give the domain of each of the following rules. Use inequality notation.

1. $F(x) = \sqrt{x - 5}$
2. $G(t) = \sqrt{t - 5}$
3. $H(y) = \sqrt{5 - y}$
4. $F(r) = \sqrt{r + 5}$
5. $G(s) = \sqrt{s^2}$
6. $H(u) = \frac{1}{u - 3}$
7. $F(v) = \frac{1}{3 - v}$
8. $G(w) = \frac{1}{3 + w}$
9. $H(x) = \frac{1}{x^2}$
10. $F(y) = \frac{1}{y^2 + 1}$
11. $G(z) = \frac{1}{z^2 - 1}$
12. $H(r) = \frac{1}{\sqrt{r}}$
13. $F(s) = \frac{1}{\sqrt{s - 5}}$
14. $G(t) = \frac{1}{\sqrt{5 - t}}$
15. $H(u) = \frac{1}{\sqrt{u + 5}}$
16. $F(v) = \frac{1}{v^2 - 9}$
17. $G(w) = \frac{(8 - w)(w - 3)}{w - 3}$
18. $H(x) = \frac{x^3}{x}$

ANSWERS TO EXERCISES

(1) $F(x) = \sqrt{x-5}$, $x \geq 5$

(2) $G(t) = \sqrt{t-5}$, $t \geq 5$

(3) $H(y) = \sqrt{5-y}$, $y \leq 5$

(4) $F(r) = \sqrt{r+5}$, $r \geq -5$

(5) $G(s) = \sqrt{s^2}$, all real numbers

(6) $H(u) = \frac{1}{u-3}$, $u \neq 3$

(7) $F(v) = \frac{1}{3-v}$, $v \neq 3$

(8) $G(w) = \frac{1}{3+w}$, $w \neq -3$

(9) $H(x) = \frac{1}{x^2}$, $x \neq 0$

(10) $F(y) = \frac{1}{y^2+1}$, all real numbers

(11) $G(z) = \frac{1}{z^2-1}$, $z \neq -1, 1$
<u>or</u> $z < -1$, $-1 < z < 1$, $z > 1$

(12) $H(r) = \frac{1}{\sqrt{r}}$, $r > 0$

(13) $F(s) = \frac{1}{\sqrt{s-5}}$, $s > 5$

(14) $G(t) = \frac{1}{\sqrt{5-t}}$, $t < 5$

(15) $H(u) = \frac{1}{\sqrt{u+5}}$, $u > -5$

(16) $F(v) = \frac{1}{v^2-9}$, $v \neq -3, 3$
<u>or</u> $v < -3$, $-3 < v < 3$, $v > 3$

(17) $G(w) = \frac{(8-w)(w-3)}{w-3}$, $w \neq 3$
<u>or</u> $w < 3$, $w > 3$

(18) $H(x) = \frac{x^3}{x}$, $x \neq 0$
<u>or</u> $x < 0$, $x > 0$

ADDITIONAL EXERCISES

Give the domain of each of the following rules. Use inequality notation.

1. $F(x) = \sqrt{x - 4}$
2. $G(t) = \sqrt{t - 4}$
3. $H(y) = \sqrt{4 - y}$
4. $F(r) = \sqrt{r + 4}$
5. $G(s) = \sqrt{s^2 + 1}$
6. $H(u) = \dfrac{1}{u - 4}$
7. $F(v) = \dfrac{1}{4 - v}$
8. $G(w) = \dfrac{1}{4 + v}$
9. $H(x) = \dfrac{1}{(x + 1)^2}$
10. $F(y) = \dfrac{1}{y^2 + 2}$
11. $G(z) = \dfrac{1}{z^2 - 16}$
12. $H(r) = \dfrac{1}{\sqrt{2r}}$
13. $F(s) = \dfrac{1}{\sqrt{s - 6}}$
14. $G(t) = \dfrac{1}{\sqrt{6 - t}}$
15. $H(u) = \dfrac{1}{\sqrt{u + 6}}$
16. $F(v) = \dfrac{1}{v^2 - 25}$
17. $G(w) = \dfrac{(w - 5)(w - 6)}{w - 6}$
18. $H(x) = \dfrac{x^2}{x}$

SUMMARY OF LESSON 5

The domain of a rule is the set of all inputs to the rule that give outputs [page 82]. For now, there are only two things to watch out for in determining the domain of a rule: no negative under a square root sign, and no zero in any denominator [page 84]. If we already have the graph of a rule, we can determine the domain by looking at the graph: the inputs that lie directly above or below the curve are the inputs that make up the domain [page 84]. We can shade this part of the x-axis to show the domain, using a large solid dot •—— at the end if the end point is included [page 86], and a large hollow dot ○—— if the end point is not included.

There is a mathematical shorthand for specifying the domain of a rule. (This shorthand also has many other uses, as we will see.) It is called "inequality notation." It uses the following symbols:

$<$ (less than), $>$ (greater than), $\leqslant$ (less than or equal to), $\geqslant$ (greater than or equal to), and $\neq$ (not equal to). The use of these symbols is explained on pages 86-88, and they are used to specify the domains of rules on page 89.

If you have trouble with any of the exercises, and the answers don't provide enough help, use the following as a guide for review.

exercises	review
1-4	example 4, page 83; example 6, page 84; example 12, page 89
5	page 84, top
6-8	example 1, page 82; example 13, page 89
9-11, 16	example 2, page 82; example 13, page 89
12-15	example 5, page 83; example 7, page 85; example 12, page 89
17-18	example 14, page 89

Lesson 6

RESTRICTED DOMAINS, FUNCTIONS, AND EQUATIONS

RESTRICTED DOMAINS

EXAMPLE 1 Example 4 of lesson 3 explained a method of cutting and folding a 12-by-10-inch cardboard sheet to make a box. We were interested in the volume of the box, and we found a rule to describe this: $V(x) = x(12 - 2x)(10 - 2x)$, where the input is the length of each cut and the output is the volume.

Problem: Write an inequality that restricts the inputs of this rule to ones that make sense in the realistic situation.

We discussed this before on pages 44-45. A negative input doesn't make sense in this situation, since the length of the cut cannot be negative. Also, cut-lengths larger than 5 don't make sense, since the cardboard sheets are only 10 inches wide. So $\boxed{0 \leqslant x \leqslant 5}$ are the only inputs that make sense in the realistic situation.

It would be convenient if we could use $0 \leqslant x \leqslant 5$ as the domain for rule V. There is a special notation to indicate this. We write the rule as follows:

$$\text{rule V:} \quad V(x) = x(12 - 2x)(10 - 2x), \; 0 \leqslant x \leqslant 5.$$

When a formula is followed by an inequality, then the inequality specifies the domain for the rule.

Written this way, rule V is a more complete description of the box-making situation than it was without the inequality. We can tell at a glance what inputs are possible in the situation at hand.

Note: The domain of rule V, as written above, is $0 \leqslant x \leqslant 5$. There is no way to figure out this domain just by looking at the formula $x(12 - 2x)(10 - 2x)$. Any input to this formula gives an output. We have to look at the inequality following the formula to see that domain has been restricted to fit the situation at hand.

To avoid confusion, mathematicians use the following convention:

1. When a formula is written without an inequality after it, then the domain of the rule is all inputs that give outputs. This is often called the natural domain.
2. When a formula is followed by an inequality, then the domain of the rule is given by the inequality. This is sometimes called a restricted domain.

EXAMPLE 2 What is the domain of the following rule?

rule V: $V(x) = x(12 - 2x)(10 - 2x)$, $0 \leqslant x \leqslant 5$

Answer: $0 \leqslant x \leqslant 5$

EXAMPLE 3 What is the domain of the following rule?

rule F: $F(x) = x(12 - 2x)(10 - 2x)$

Since there is no restriction on the domain, we give the natural domain. In this case, it is all real numbers.

Answer: All real numbers

EXAMPLE 4 In each case, give the domain of the rule.

rule	domain
rule F: $F(x) = \frac{1}{x}$, $x > 0$	$x > 0$
rule G: $G(x) = \frac{1}{x}$	$x \neq 0$
rule F: $F(x) = \sqrt{x + 4}$	$x \geqslant -4$
rule G: $G(x) = \sqrt{x + 4}$, $5 < x < 21$	$5 < x < 21$

In the second and third cases we gave the natural domain, while in the first and fourth cases we repeated the inequality.

EXAMPLE 5 What is the domain of the following rule?

rule F: $F(x) = \frac{1}{x}$, $x \neq 0$

Answer: $x \neq 0$. Note that this is also the natural domain of $\frac{1}{x}$. Some math books make a practice of always writing rules this way, even to indicate the natural domain.

Note: You aren't completely free to specify any domain you want simply by putting an inequality after a formula. For instance, it would be a mistake to write $F(x) = \sqrt{x}$, $-2 < x < 2$. This can't be the domain of F, since $-2 < x < 2$ includes negative inputs, but the rule is impossible to follow with negative inputs. A restricted domain can't include numbers that are not part of the natural domain.

Let's see how to graph a rule with a restricted domain.

<u>EXAMPLE 6</u> Graph the following rules.

rule F: $F(x) = x(12 - 2x)(10 - 2x)$

rule V: $V(x) = x(12 - 2x)(10 - 2x)$, $0 \leqslant x \leqslant 5$

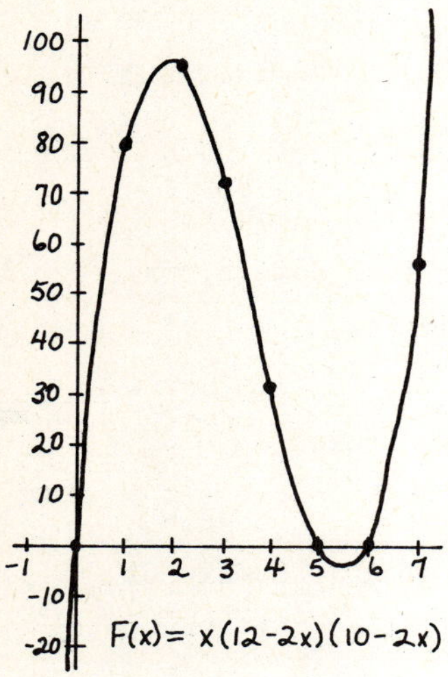

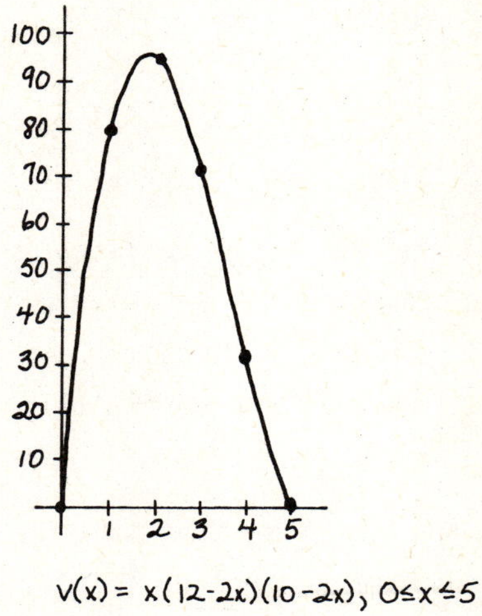

(We saw these graphs before in example 5 of lesson 3.)

<u>Note:</u> 1. It was actually easier to graph rule V, since there was no question about what part of the graph we were supposed to display: the part with inputs running from 0 to 5. To graph rule F, we had to decide for ourselves which part of the graph was the best part to display. We couldn't draw the whole thing, since the curve never ends.

2. The graph of rule V has definite <u>end points</u>, one for the input 0 and one for the input 5. We indicated these with big, solid dots. The graph of rule F, on the other hand, goes on forever. We indicated this by drawing the curve to the edge of the picture, and leaving it with no dots on its ends.

<u>Here are some conventions about drawing end points of curves.</u> Let's start with a rule we've seen before:

EXAMPLE 7

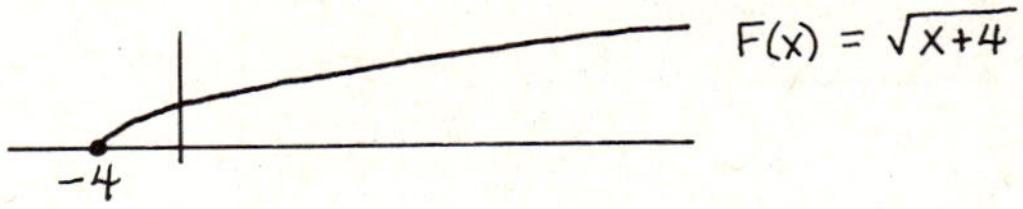

The natural domain of the rule is $x \geqslant -4$. Notice that -4 is included in the domain, so that there is an (input,output) pair to plot for this input. We have indicated this with a big, solid dot. But to the right, the graph goes on without limit. We have indicated this by leaving the curve with no dot on its end.

Now let's restrict the domain of this rule in various ways, and see how the graph is affected.

EXAMPLE 8

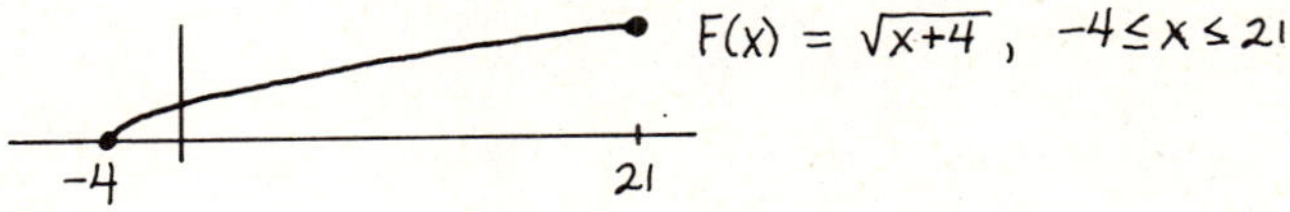

Solid dots indicate the end points.

EXAMPLE 9

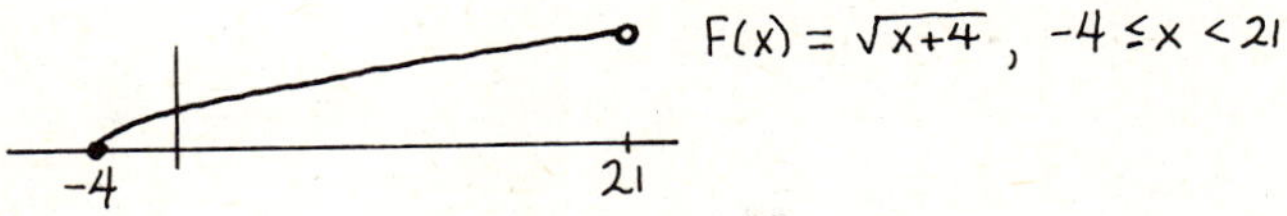

21 is not included in $-4 \leqslant x < 21$, so there is no (input,output) pair for an input of 21. The curve comes right up to 21, but does not include it. So the curve is just like the one in example 8 except that it doesn't include the right-hand end point. We indicate this by putting a big, hollow dot in place of the solid one. (If we didn't use any dot, then it would look as if the curve goes on forever, as in example 7.)

EXAMPLE 10

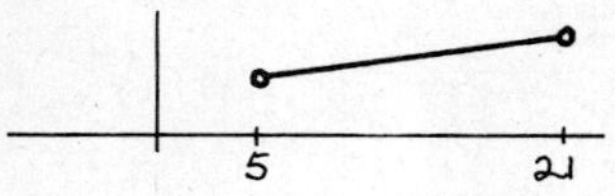

EXAMPLE 11

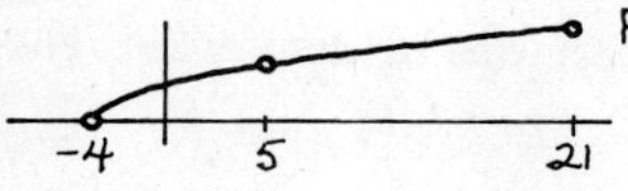

This time the domain is given by $-4 < x < 21$, $x \neq 5$. This excludes -4, 5, and 21. Even though the point above 5 is not an end point, we show that it has been excluded by replacing it with a big, hollow dot.

EXAMPLE 12 Graph $G(x) = \frac{(x + 1)(x - 1)}{(x - 1)}$, and $F(x) = x + 1$. Notice the difference in the graphs.

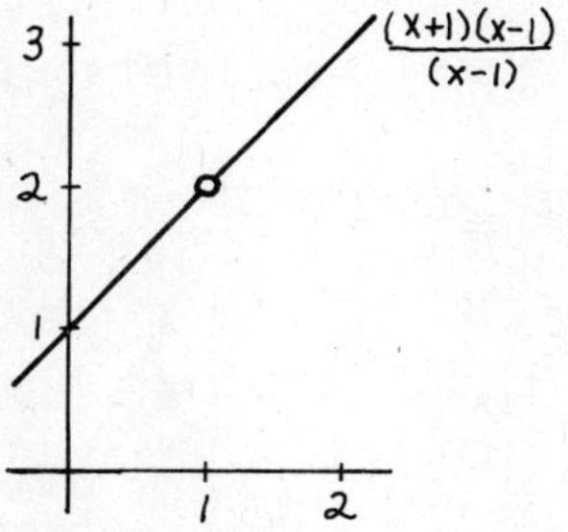

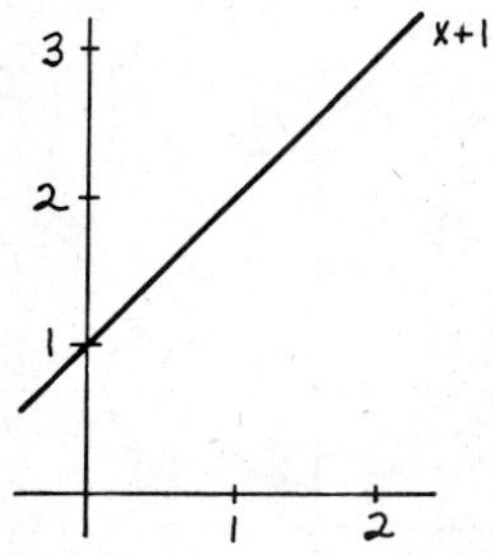

We saw on pages 89-90 that the natural domain of $\frac{(x + 1)(x - 1)}{(x - 1)}$ is $x \neq 1$. Consequently, there is no (input,output) pair above the input 1 on the graph of this formula. We indicate this with a hollow dot. Otherwise, the graphs of F and G are the same.

Notice that the denominator of $\frac{(x + 1)(x - 1)}{(x - 1)}$ is zero when the input is 1, but the graph does not have a vertical asymptote there. This is the first such example we have seen. Check the graph carefully using inputs close to 1 to see that this is true.

Here is an algebraic way to see why this is true. If the numerator and denominator of a fraction have equal factors, we can cancel these factors, as long as they aren't zero. So as long as the denominator of $\frac{(x + 1)(x - 1)}{(x - 1)}$ is not zero, its output is equal to x + 1.

So the graph of $\frac{(x + 1)(x - 1)}{(x - 1)}$ is the same as the graph of x + 1, except that the point above 1 is missing on the first graph. The graph of x + 1 is a straight line. It is not a curve with an asymptote.

Summary

✓ Whenever we want to indicate whether or not a certain (input,output) pair should be included on a curve, we use a solid dot to indicate that it should be, or a hollow dot to indicate that it should not.

✓ Whenever we want to indicate that a curve goes on forever, we draw it to the edge of the picture and leave it with no dot at the end.

✓ If the denominator in a formula has a zero, then the rule is undefined there, and there will be no (input,output) pair on the graph. There are two possibilities for how the graph will look:

If the denominator can be cancelled with an equal factor in the numerator, as in example 12, then there will be a "hole" in the curve over the bad input 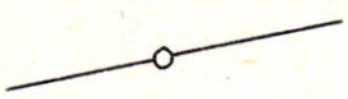 . You can check this with inputs close to the "bad" one.

Otherwise, there will be a vertical asymptote through the bad input. You can also check this with inputs close to the "bad" one.

Seg. 9: Do exercises 1-3, p. 111, then start next Lesson.

GRAPHING SPLIT FORMULAS

In lesson 4 of Module X, we saw that certain rules use more than one formula. For example:

$$\text{rule F:} \quad F(x) = \begin{cases} 4, & \text{if the input is 3 or less} \\ 5 - x, & \text{if the input is more than 3} \end{cases}$$

$$\text{or} \quad F(x) = \begin{cases} 4, & x \leqslant 3 \\ 5 - x, & x > 3 \end{cases}$$

EXAMPLE 13 Graph rule F.

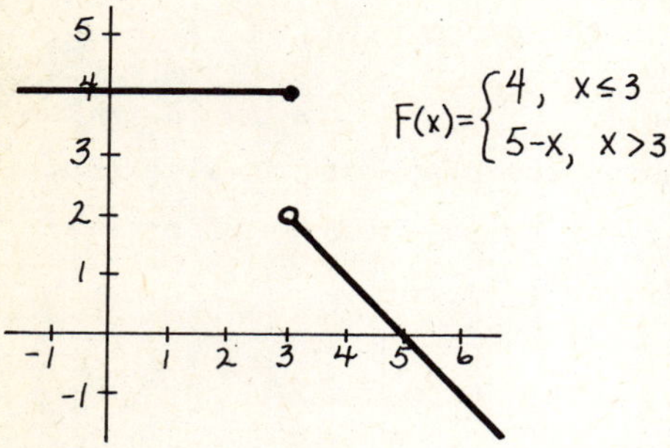

We simply graphed both parts on the same picture. Notice that according to the formula, $F(3) = 4$. This is because the first part has the domain $x \leqslant 3$, which includes 3. On the graph, this is indicated by the big, solid dot at the right end of the left-hand part. If we start at 3 on the x-axis and go straight up until we meet the curve, there is no question about where we meet the curve—at the solid dot, not the hollow one.

EXAMPLE 14 Graph $G(x) = \begin{cases}4, & x < 3\\ 5 - x, & x \geqslant 3\end{cases}$.

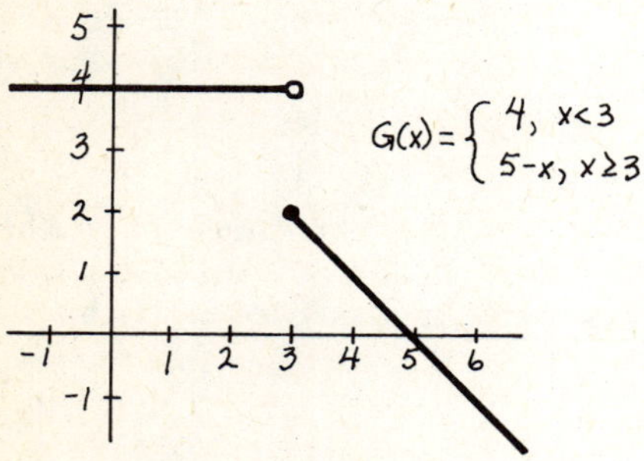

Notice the difference between this example and the last one.

If we tried to draw these graphs without the solid and hollow dots, the graph would be ambiguous:

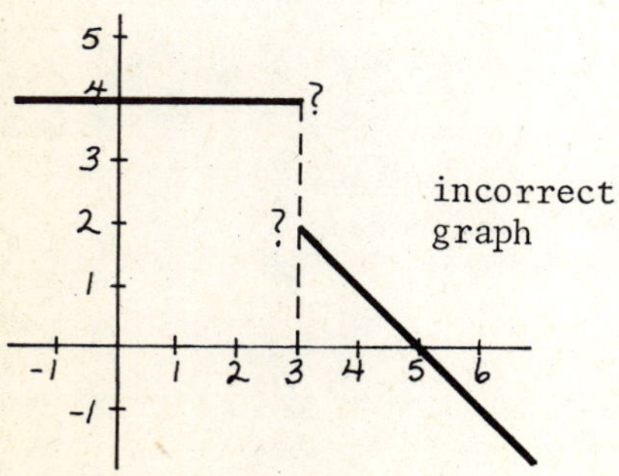

We can't tell from this graph which output goes with input 3.

EXAMPLE 15 Graph $H(x) = \begin{cases} 2, & x \leqslant 3 \\ 5 - x, & x > 3 \end{cases}$.

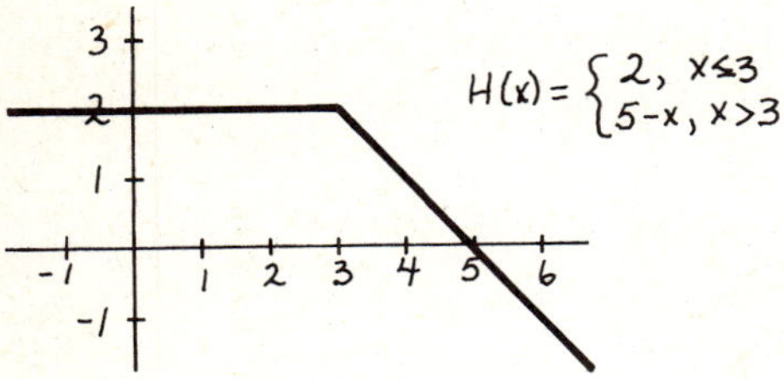

This time we don't need to use solid and hollow dots. Since the two parts meet, there is no question on this graph about what output goes with input 3. We can see that H(3) = 2.

EXAMPLE 16 Here is a rule that is often used in programming computers. It is called the "sign" rule. (It has no relation to the "sine" rule in trigonometry, even though they sound the same.) The sign rule simply gives an output of 1, 0, or -1 depending on whether the input is positive, zero, or negative.

$$\text{rule S:} \quad S(x) = \begin{cases} 1, & x > 0 \\ 0, & x = 0 \\ -1, & x < 0 \end{cases}$$

Notice that the input 0 is included only with the middle part, not with the first or third parts. Let's graph this rule:

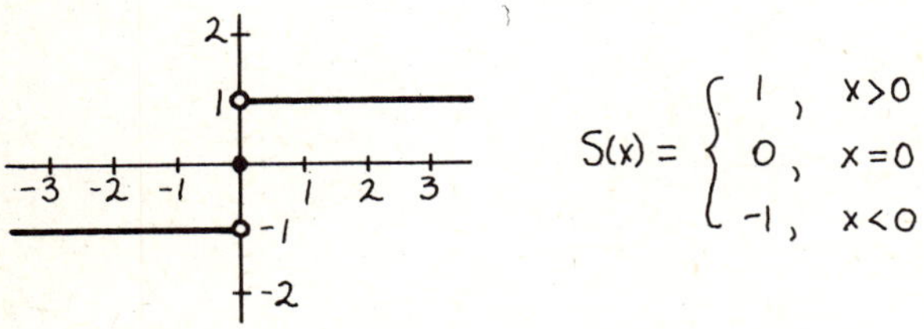

When the input is 0, the graph shows an output of 0. When the input is greater than 0, the graph shows an output of 1. And when the input is less than 0, the graph shows an output of -1.

EXAMPLE 17 Graph $A(x) = |x|$. This is the absolute value rule that we discussed in Module X. We could simply plot points, but let's use the three-part formula for this rule that we gave in lesson 4, Module X.

$$\text{rule A:} \quad A(x) = \begin{cases} x, & x > 0 \\ 0, & x = 0 \\ -x, & x < 0 \end{cases}$$

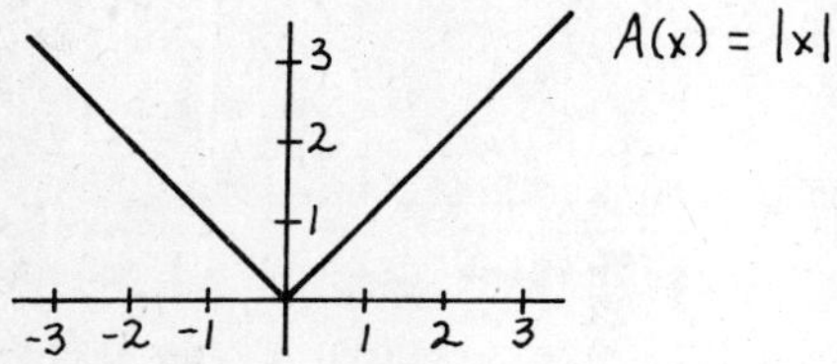

Notice that in this case, there is no need to use hollow or solid dots. Since all three parts come together, there is never any question about the correct output.

Note: The graphs of multiple-part rules that we have been looking at in examples 13-17 have some new features that we have not seen before. Some of them have sharp corners , and some of them have "jumps" 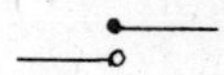. None of the graphs we have seen before this have had sharp corners, and the only kind of "jumps" we have seen have been at vertical asymptotes 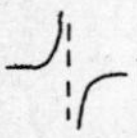. Keep this in mind when you are checking your graphs. If you are graphing a single-formula rule and you draw a sharp point , or a jump without a vertical asymptote, then you have probably made a mistake.

THE VERTICAL LINE TEST

Here is a new kind of graph:

EXAMPLE 18

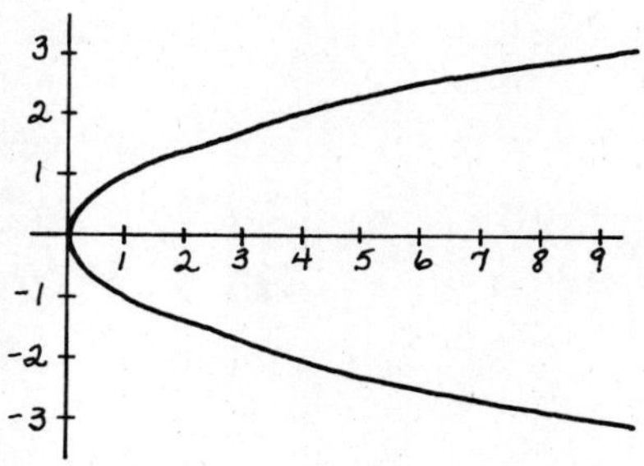

The thing that is new about this graph is that it shows more than one output for each input. For instance:

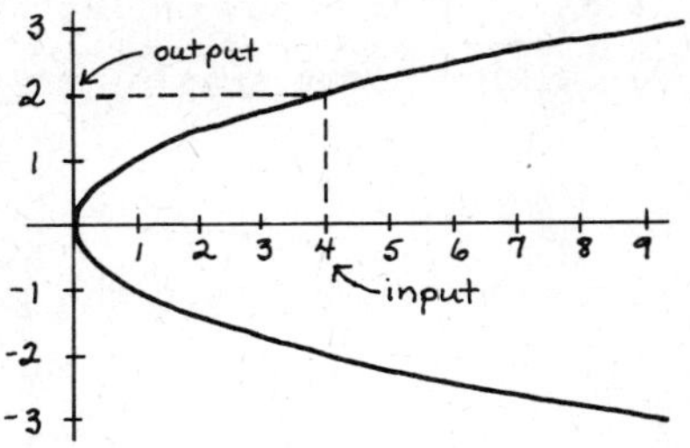

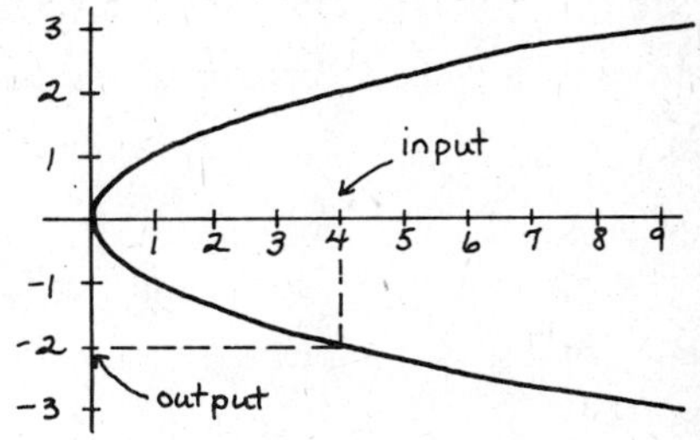

For an input of 4, the graph shows two different outputs, 2 and -2.

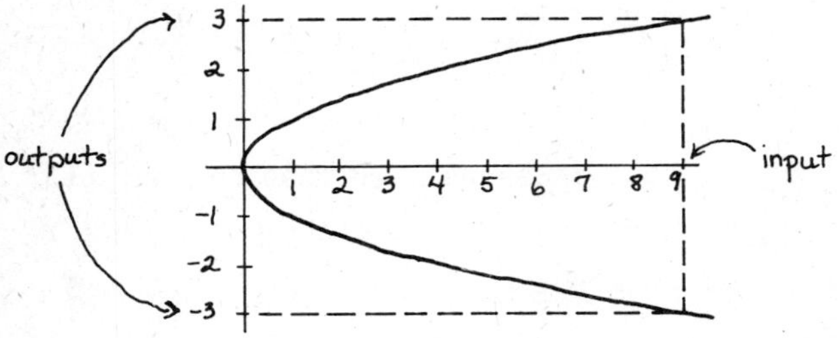

For an input of 9, the graph shows outputs of 3 and -3.

What kind of input-output rule could have a graph like this? So far, every rule that we have seen has told us how to take an input and compute a single output. The following rule is different.

rule Z: Take the input and compute its square root. Give both the positive square root and the negative square root as outputs.

The graph above is a graph of rule Z. For each input, the rule gives both the positive and negative square roots as outputs.

Rules such as Z above are not very common in mathematics. Mathematicians are mainly interested in rules that give only a single output for each input. There is a special word for this:

A function is a rule that gives at most one output for each input.

So rule Z above is not a function, but every other input-output rule we have seen up until now is a function. And probably every rule you will see in other math courses will be a function. So from now on, we will often use the word "function" when we are talking about single-output rules.

Every graph we have seen, up until the last example, has been the graph of a function. As example 18 showed, there is an easy way to tell whether or not a graph represents a function. Simply look to see whether or not it ever shows more than one output for the same input. If it does, then it is not the graph of a function.

EXAMPLE 19 Is the following the graph of a function?

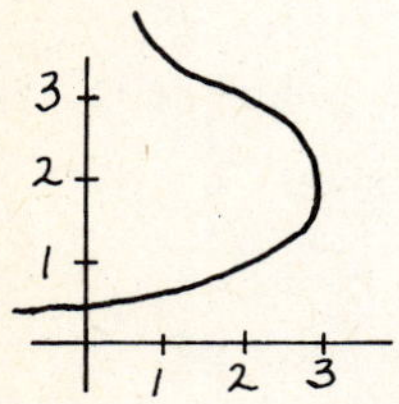

Answer: No, because it shows more than one output for some inputs. For instance, it shows two different outputs for an input of 2.

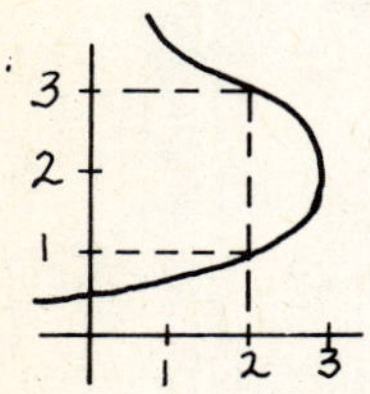

Note: The vertical dotted line in the picture to the left meets the curve at two different points.

Whenever it is possible to draw a vertical line that meets a curve more than once, then the curve cannot be the graph of a function.

This is called the VERTICAL LINE TEST.

EXAMPLE 20 In each case, use the vertical line test to see whether or not the curve is the graph of a function.

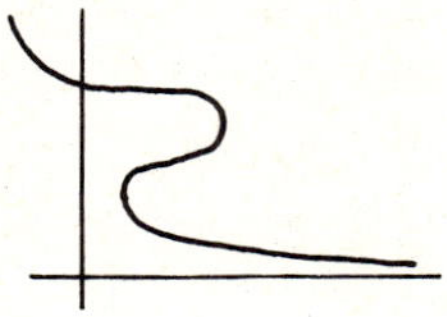

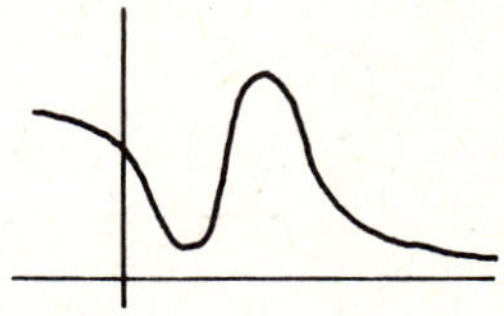

Answers:

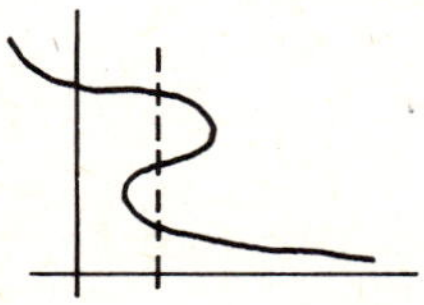

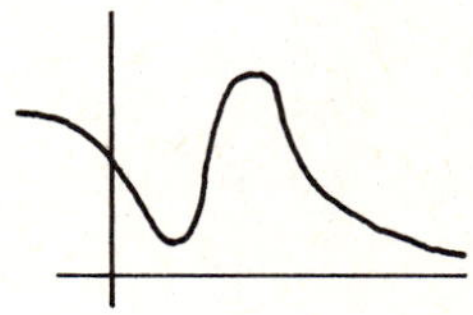

not a function

This one fails the vertical line test.

function

This one passes the vertical line test, since out of all the possible vertical lines, none meets the curve more than once.

One use for the vertical line test is in checking your graphs. If you are graphing a rule that gives only one output for each input (as is almost always the case), then your graph should pass the vertical line test. If it doesn't, you have probably made a mistake. For instance, if you draw a parabola like this

, it will fail the vertical line test: 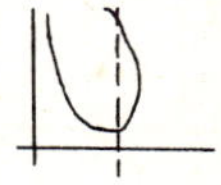.

GRAPHING EQUATIONS

So far, we have concentrated on graphing functions, such as $F(x) = x^2 - 5x + 4$. But you probably have also had experience with graphing equations, such as $y = x^2 - 5x + 4$. In this case, the function and the equation are practically identical, and as we pointed out in lesson 1, the graphs of both are exactly the same. But equations can look different from formulas, and their graphs may not look like graphs of functions (that is, they may not pass the vertical line test.)

<u>EXAMPLE 21</u> Graph $x - 2y = 2$. Written this way, it doesn't look like an input-output rule. But we can make it look like one by getting y alone on the left-hand side of the equation.

$$\begin{array}{rcl} x - 2y & = & 2 \\ \underline{-x \qquad} & & \underline{-x} \\ \dfrac{-2y}{-2} & = & \dfrac{-x + 2}{-2} \end{array}$$

$$y = \frac{-x + 2}{-2} = \frac{-x}{-2} + \frac{2}{-2} = \frac{1}{2}x - 1$$

Now we can graph $y = \frac{1}{2}x - 1$, just as we could graph $F(x) = \frac{1}{2}x - 1$.

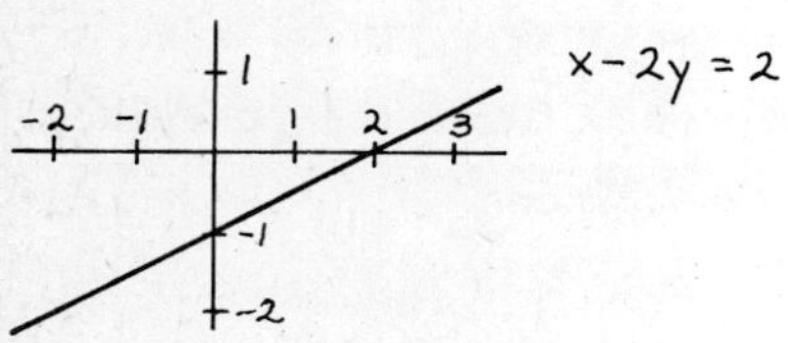

$x - 2y = 2$ was a linear equation. We put y alone on the left, and graphed it in the familiar way. The graph is identical to the graph of $F(x) = \frac{1}{2}x - 1$, and it passes the vertical line test.

<u>EXAMPLE 22</u> Graph $y^2 = x$. Again, we begin by getting y alone on the left-hand side. To do this, we have to "unsquare" both sides:

$$y^2 = x$$

$$y = \begin{cases} \sqrt{x} \\ -\sqrt{x} \end{cases}$$

Notice that we have to give both the positive <u>and</u> negative square root as values for y, or we would be leaving out some solutions. (For instance, if we put in 9 for x, then both 3 <u>and</u> -3 will work for y.)

We will graph this the same way that we have graphed other two-part rules. Since the domain hasn't been specified, we will use the natural domain of each part. In this case, both domains are the <u>same</u>, $x \geqslant 0$.

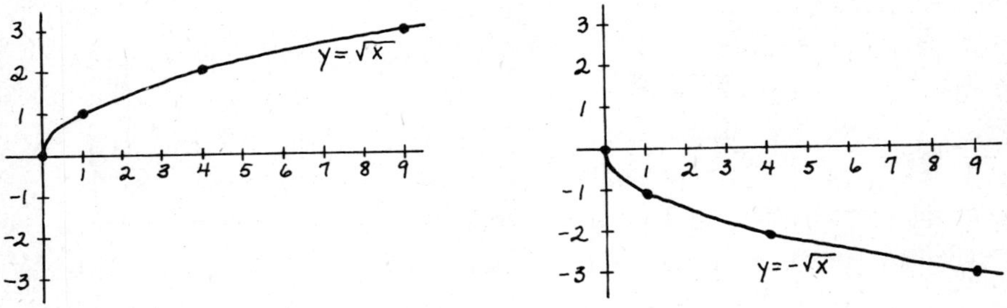

Now we combine these graphs into one picture. (In doing this yourself, you would draw them both on the same picture in the first place.)

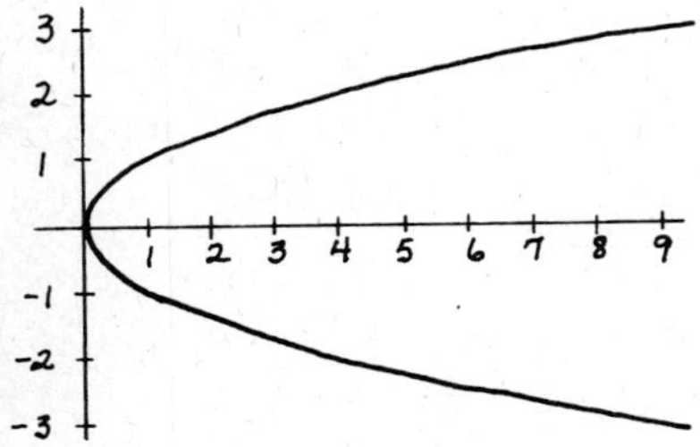

$y^2 = x$, or $y = \begin{cases} \sqrt{x} \\ -\sqrt{x} \end{cases}$

Notice that this is the same as rule Z in example 17. The graph doesn't pass the vertical line test.

The last example illustrates the following procedure:

> To graph an equation, first get y alone on the left. In doing this, one of the steps may be to unsquare both sides. When you do this, you have to split the right-hand side into two parts, one for the positive square root and one for the negative square root. Then graph both parts on the same picture.

EXAMPLE 23 Graph $x^2 + y^2 = 25$.

First, we get y alone on the left.

$$x^2 + y^2 = 25$$
$$-x^2 \qquad -x^2$$
$$y^2 = 25 - x^2$$
$$y = \begin{cases} \sqrt{25 - x^2} \\ -\sqrt{25 - x^2} \end{cases}$$

Then we graph both parts on the same picture. (If you are following the advice given at the beginning of this module and reading actively, then you should try to graph this for yourself before reading on. Be sure to try negative inputs, as well as positive ones.)

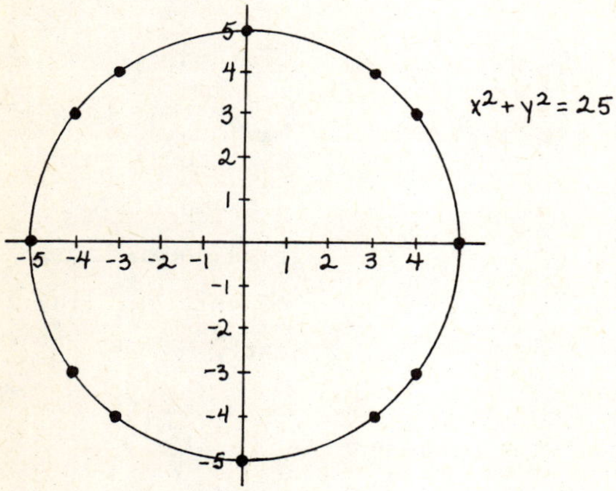

The top half is the graph of $y = \sqrt{25 - x^2}$, and the bottom half is the graph of $y = -\sqrt{25 - x^2}$. Combined, they are the graph of $y = \begin{cases} \sqrt{25 - x^2} \\ -\sqrt{25 - x^2} \end{cases}$, which is the same as $x^2 + y^2 = 25$.

This is a circle of radius 5. Circles and their equations are studied in detail in many calculus courses.

EXERCISES (Answers begin on page 113.)

Graph each function.

1. $F(x) = 8 - x, x \geqslant 3$
   $G(x) = 8 - x, x > 3$
   $H(x) = 8 - x, x \neq 3$
   $J(x) = 8 - x, 3 < x \leqslant 10$

2. $F(x) = x^2, x > 0$
   $G(x) = x^2, x \leqslant 0$
   $H(x) = x^2, x \neq 0$
   $J(x) = x^2, -1 \leqslant x < 1$

3. $F(x) = x(x + 1)(x - 1), -1 \leqslant x \leqslant 1$
   $G(x) = x(x + 1)(x - 1), x < -1$
   $H(x) = x(x + 1)(x - 1), x \leqslant 0$
   $J(x) = x(x + 1)(x - 1), x \neq 0$

4. $F(x) = \begin{cases} 5, & x < 3 \\ 8 - x, & x \geqslant 3 \end{cases}$ $G(x) = \begin{cases} 2, & x < 3 \\ 8 - x, & x \geqslant 3 \end{cases}$

   $H(x) = \begin{cases} 2, & x \leqslant 3 \\ 8 - x, & x > 3 \end{cases}$ $J(x) = \begin{cases} 2, & x < 3 \\ 4, & x = 3 \\ 8 - x, & x > 3 \end{cases}$

5. $F(x) = \begin{cases} x^2, & x > 0 \\ -x^2, & x \leqslant 0 \end{cases}$ $G(x) = \begin{cases} x^2, & x > 0 \\ 2 - x^2, & x \leqslant 0 \end{cases}$

   $H(x) = \begin{cases} x^2, & x \geqslant 0 \\ 2 - x^2, & x < 0 \end{cases}$ $J(x) = \begin{cases} x^2, & x > 0 \\ 1, & x = 0 \\ 2 - x^2, & x < 0 \end{cases}$

6. $F(x) = \begin{cases} 8 - x, & x \geqslant 3 \\ 5 - x, & x < 3 \end{cases}$ $G(x) = \begin{cases} 8 - x, & x > 3 \\ 4, & x = 3 \\ 5 - x, & x < 3 \end{cases}$

   $H(x) = \begin{cases} 8 - x, & x > 3 \\ 4, & x = 3 \\ 8 - x, & x < 3 \end{cases}$ $J(x) = \begin{cases} 8 - x, & x > 3 \\ 8 - x, & x < 3 \end{cases}$

7. $F(x) = 8 - x, x \neq 3$ $G(x) = \dfrac{(8 - x)(x - 3)}{x - 3}$

8. $F(x) = x - 1, x \neq 2$ $G(x) = \dfrac{(x - 1)(x - 2)}{x - 2}$

9. $F(x) = x^2, x \neq 0$ $G(x) = \dfrac{x^3}{x}$

Graph the following equations.

10. $y^2 = x + 4$

11. $y^2 = x - 4$

12. $x^2 + y^2 = 25$

13. $(x - 3)^2 + y^2 = 25$

14. $x^2 + (y - 3)^2 = 25$

15. $9x^2 + 25y^2 = 225$

16. Which of the following are graphs of functions?

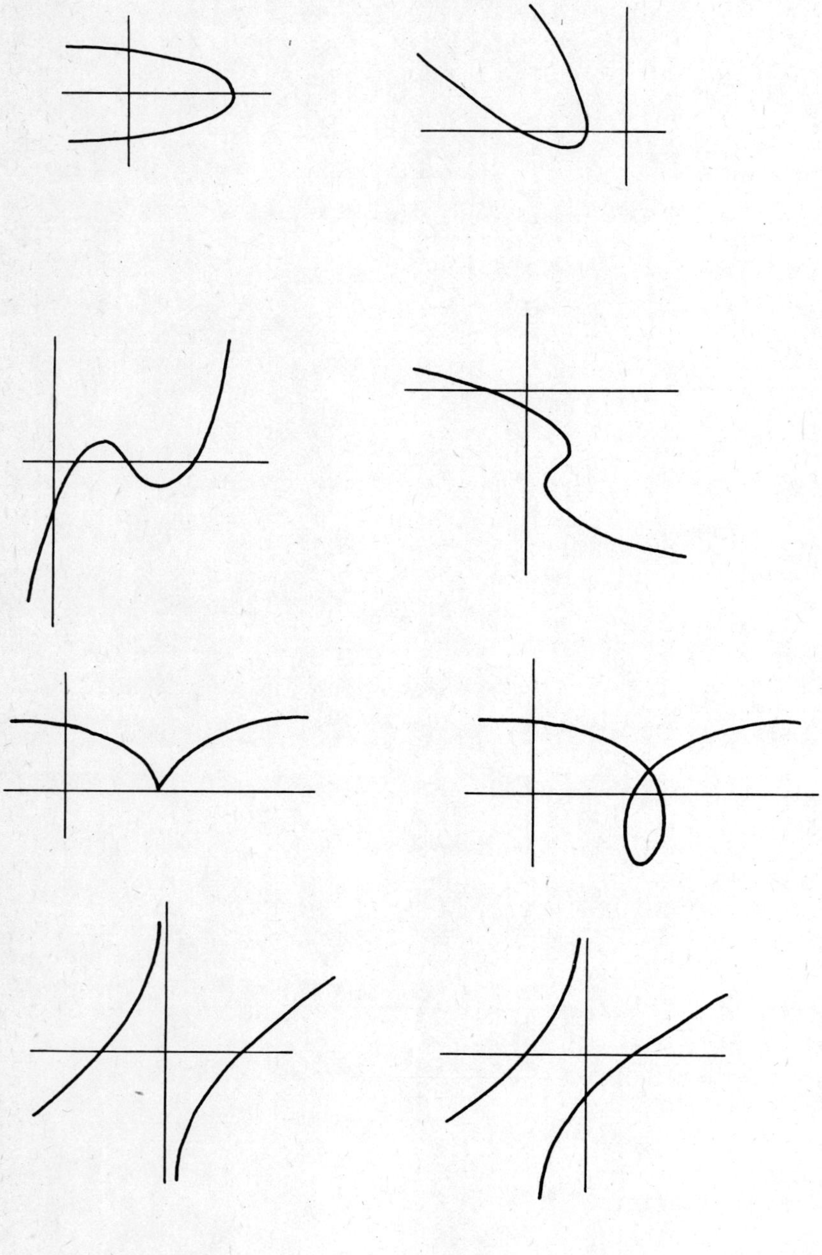

ANSWERS TO EXERCISES

(1)

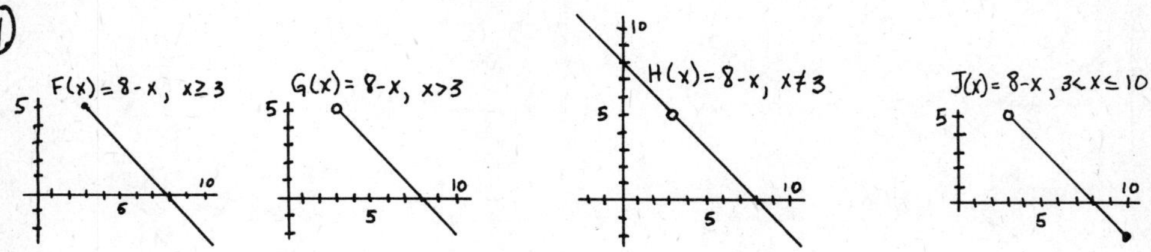

(2)

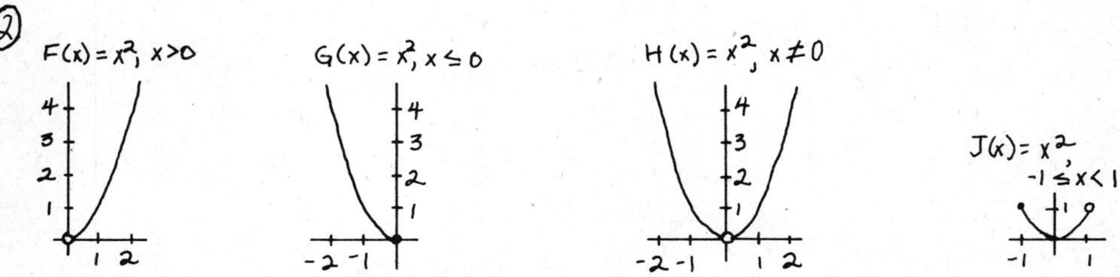

(3)

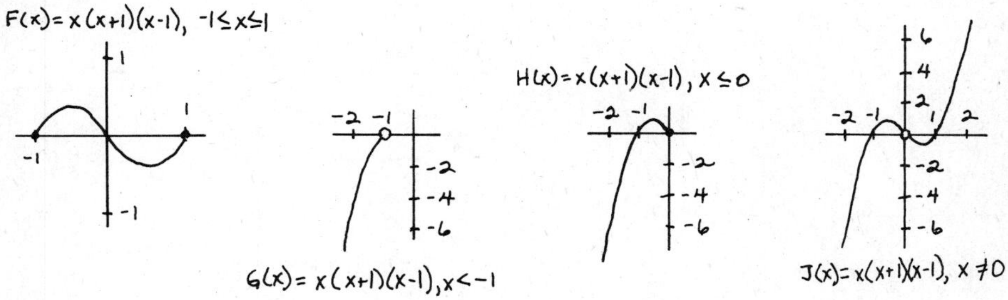

(4)

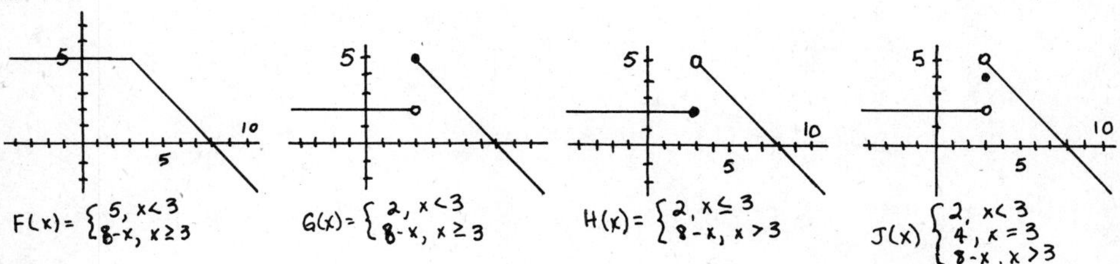

(5)

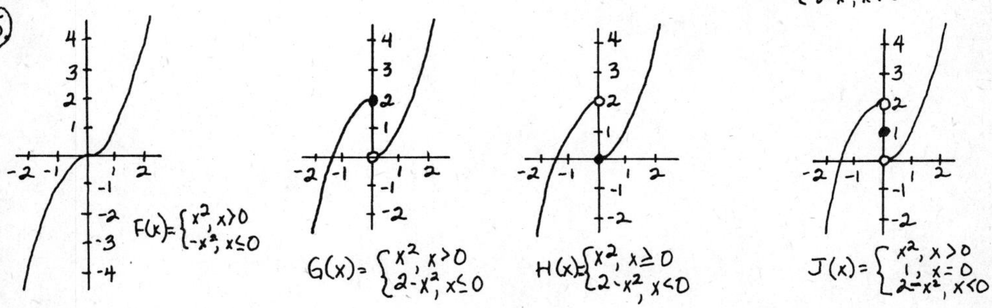

(6)

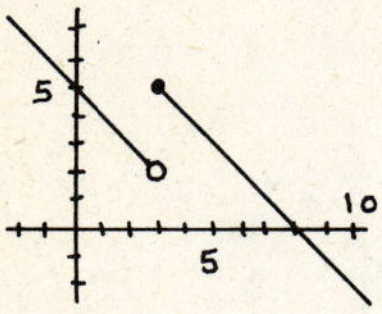

$F(x) = \begin{cases} 8-x, & x \geq 3 \\ 5-x, & x < 3 \end{cases}$

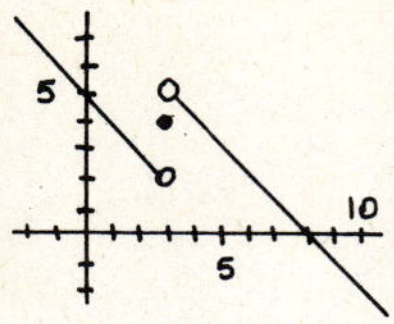

$G(x) = \begin{cases} 8-x, & x > 3 \\ 4, & x = 3 \\ 5-x, & x < 3 \end{cases}$

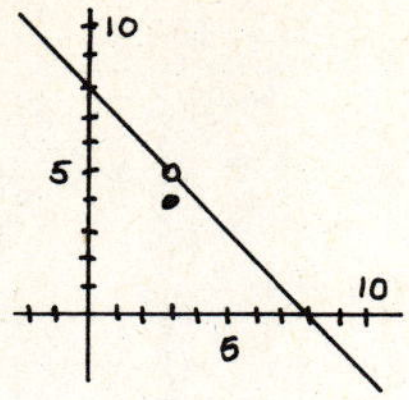

$H(x) = \begin{cases} 8-x, & x > 3 \\ 4, & x = 3 \\ 8-x, & x > 3 \end{cases}$

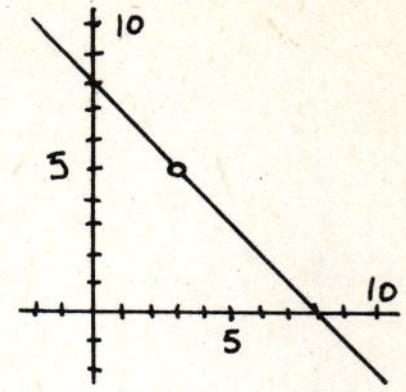

$J(x) = \begin{cases} 8-x, & x > 3 \\ 8-x, & x < 3 \end{cases}$

(7)

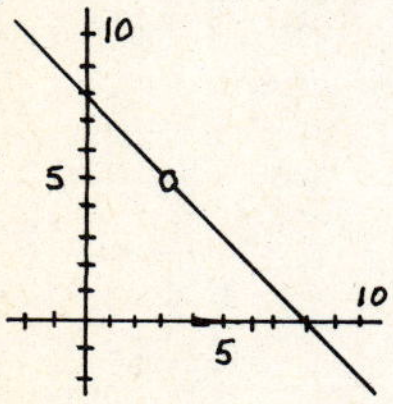

$F(x) = 8 - x, \; x \neq 3$

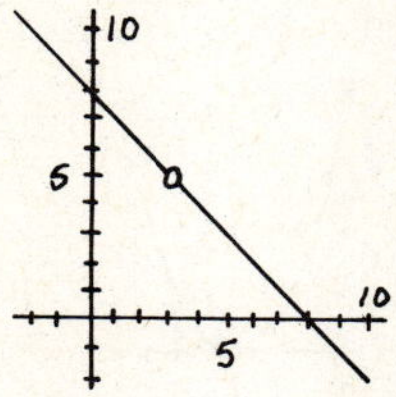

$G(x) = \dfrac{(8-x)(x-3)}{x-3}$

(8)

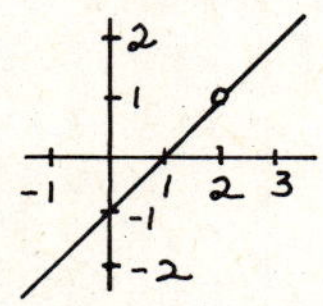

$F(x) = x - 1, \; x \neq 2$

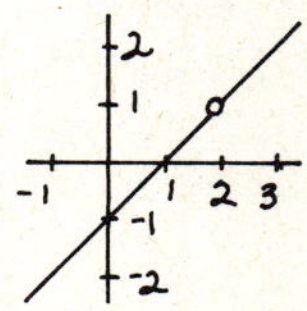

$G(x) = \dfrac{(x-1)(x-2)}{x-2}$

(9)

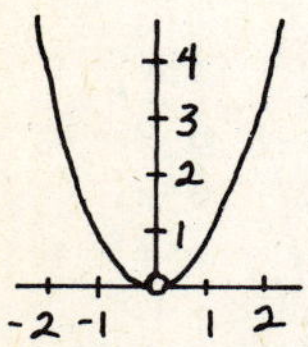

$F(x) = x^2, \; x \neq 0$

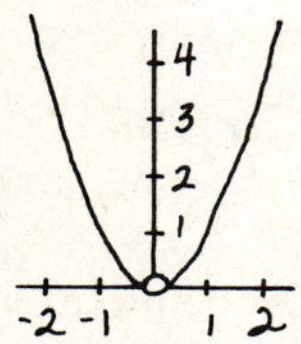

$G(x) = \dfrac{x^3}{x}$

(10)

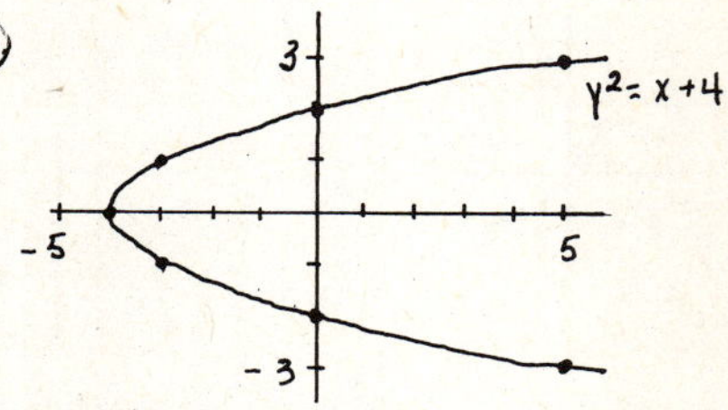

$y^2 = x + 4$

$y = \begin{cases} \sqrt{x+4} \\ -\sqrt{x+4} \end{cases}$

(11)

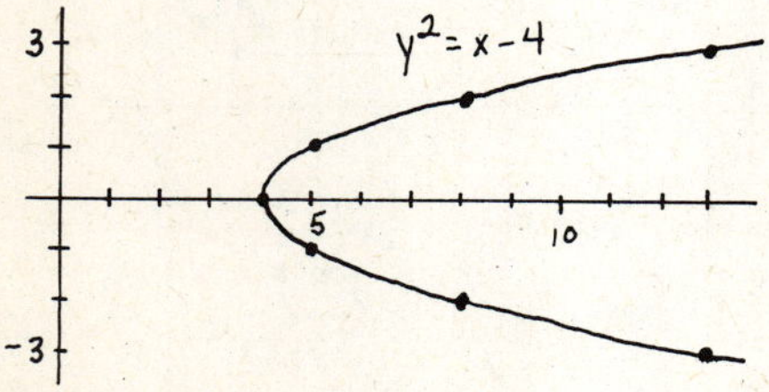

$y^2 = x - 4$

$y = \begin{cases} \sqrt{x-4} \\ -\sqrt{x-4} \end{cases}$

(12)

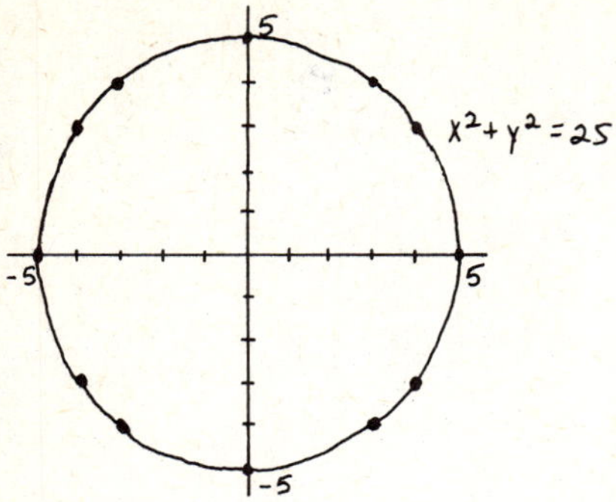

$$x^2+y^2=25$$
$$y^2=25-x^2$$
$$y^2=\begin{cases}\sqrt{25-x^2}\\-\sqrt{25-x^2}\end{cases}$$

(13)

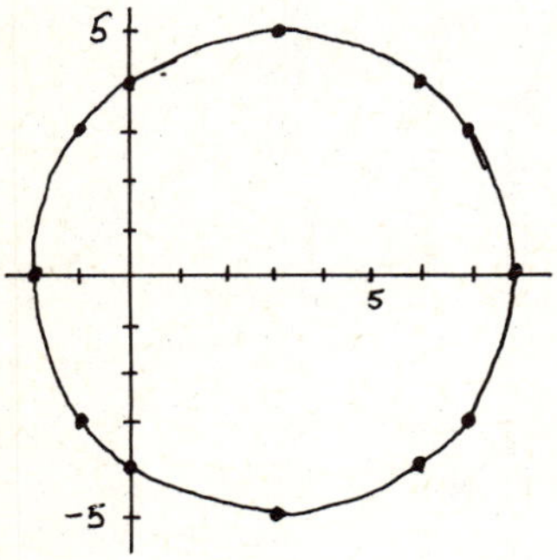

$$(x-3)^2+y^2=25$$
$$y^2=25-(x-3)^2$$
$$y=\begin{cases}\sqrt{25-(x-3)^2}\\-\sqrt{25-(x-3)^2}\end{cases}$$

(14)

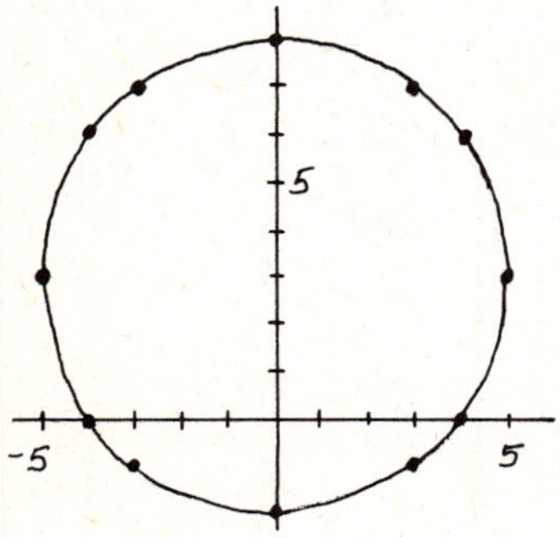

$$x^2+(y-3)^2=25$$
$$(y-3)^2=25-x^2$$
$$y-3=\begin{cases}\sqrt{25-x^2}\\-\sqrt{25-x^2}\end{cases}$$
$$y=\begin{cases}\sqrt{25-x^2}+3\\-\sqrt{25-x^2}+3\end{cases}$$

(15)

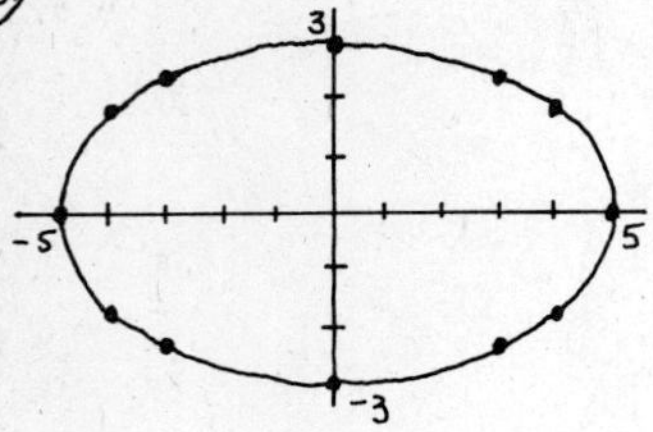

$$9x^2 + 25y^2 = 225$$

$$25y^2 = 225 - 9x^2 = 9(25 - x^2)$$

$$5y = \begin{cases} \sqrt{9(25-x^2)} \\ -\sqrt{9(25-x^2)} \end{cases}$$

$$5y = \begin{cases} 3\sqrt{25-x^2} \\ -3\sqrt{25-x^2} \end{cases}$$

$$y = \begin{cases} \frac{3}{5}\sqrt{25-x^2} \\ -\frac{3}{5}\sqrt{25-x^2} \end{cases}$$

(16)

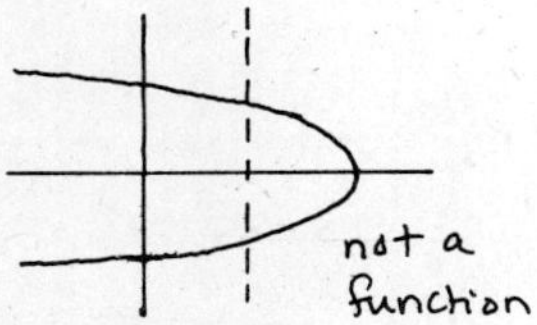

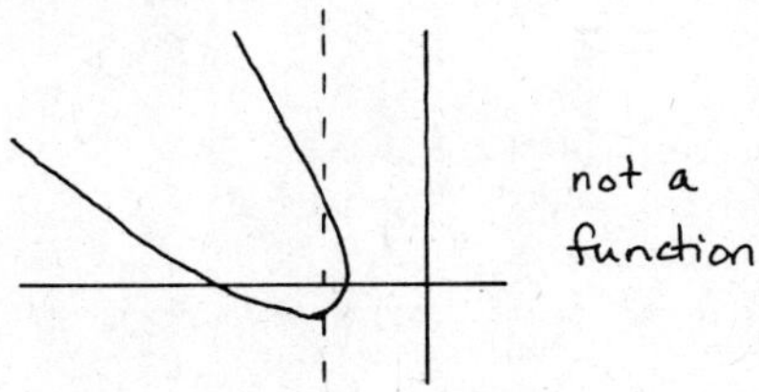

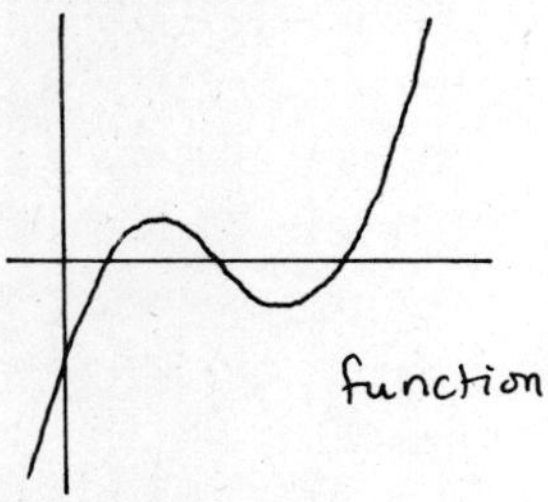

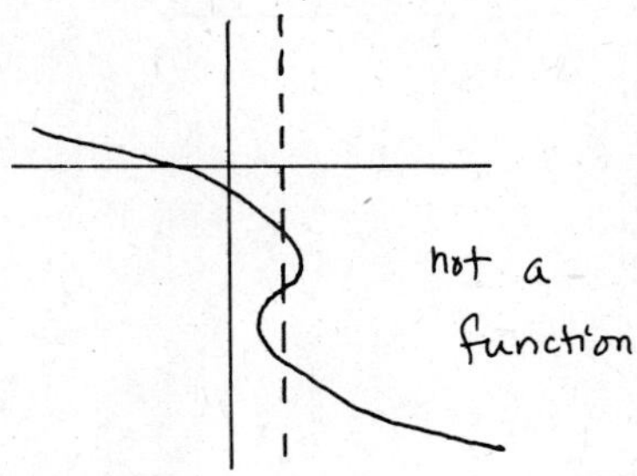

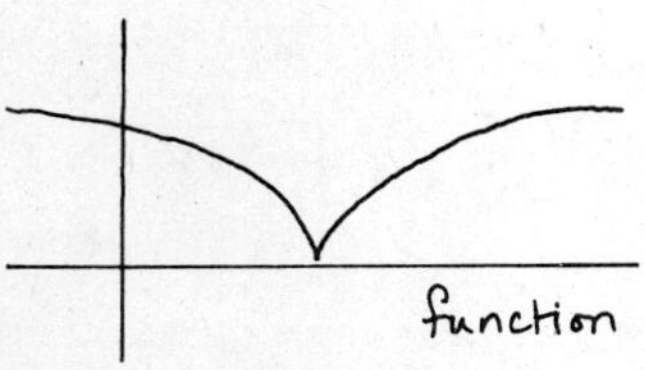

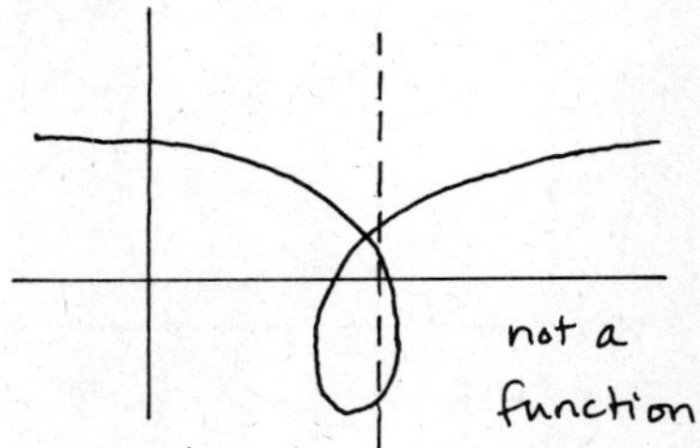

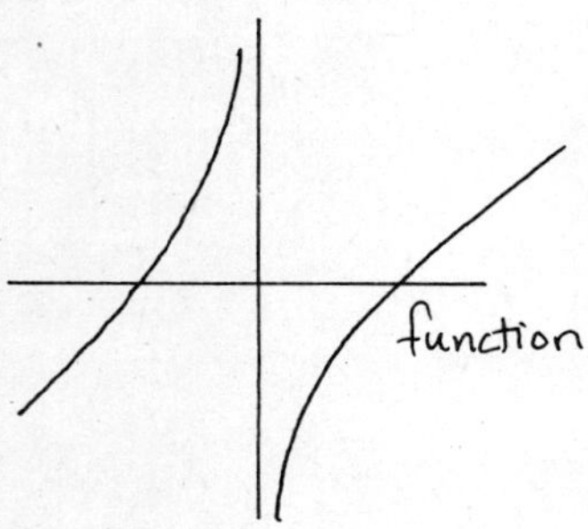

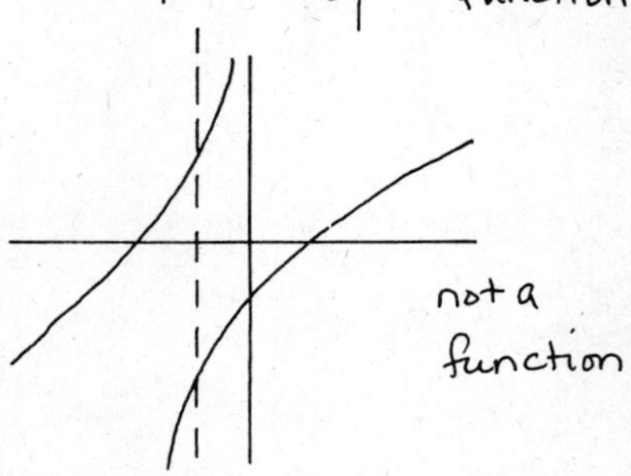

ADDITIONAL EXERCISES

Graph each function.

1. $F(x) = 7 - x,\ x \geqslant 3$
   $G(x) = 7 - x,\ x > 3$
   $H(x) = 7 - x,\ x \neq 3$
   $J(x) = 7 - x,\ 3 < x \leqslant 10$

2. $F(x) = (x + 1)^2,\ x > -1$
   $G(x) = (x + 1)^2,\ x \leqslant -1$
   $H(x) = (x + 1)^2,\ x \neq -1$
   $J(x) = (x + 1)^2,\ -2 \leqslant x < 0$

3. $F(x) = x(x - 1)(x - 2),\ 0 \leqslant x \leqslant 2$
   $G(x) = x(x - 1)(x - 2),\ x < 0$
   $H(x) = x(x - 1)(x - 2),\ x \leqslant 1$
   $J(x) = x(x - 1)(x - 2),\ x \neq 1$

4. $F(x) = \begin{cases} 4, & x < 3 \\ 7 - x, & x \geqslant 3 \end{cases}$ $G(x) = \begin{cases} 1, & x < 3 \\ 7 - x, & x \geqslant 3 \end{cases}$

   $H(x) = \begin{cases} 1, & x \leqslant 3 \\ 7 - x, & x > 3 \end{cases}$ $J(x) = \begin{cases} 1, & x < 3 \\ 2, & x = 3 \\ 7 - x, & x > 3 \end{cases}$

5. $F(x) = \begin{cases} (x + 1)^2, & x > -1 \\ -(x + 1)^2, & x \leqslant -1 \end{cases}$ $G(x) = \begin{cases} (x + 1)^2, & x > -1 \\ 2 - (x + 1)^2, & x \leqslant -1 \end{cases}$

   $H(x) = \begin{cases} (x + 1)^2, & x \geqslant -1 \\ 2 - (x + 1)^2, & x < -1 \end{cases}$ $J(x) = \begin{cases} (x + 1)^2, & x > -1 \\ 1, & x = -1 \\ 2 - (x + 1)^2, & x < -1 \end{cases}$

6. $F(x) = \begin{cases} 7 - x, & x \geqslant 3 \\ 4 - x, & x < 3 \end{cases}$ $G(x) = \begin{cases} 7 - x, & x > 3 \\ 2, & x = 3 \\ 4 - x, & x < 3 \end{cases}$

   $H(x) = \begin{cases} 7 - x, & x < 3 \\ 2, & x = 3 \\ 7 - x, & x > 3 \end{cases}$ $J(x) = \begin{cases} 7 - x, & x < 3 \\ 7 - x, & x > 3 \end{cases}$

7. $F(x) = 7 - x,\ x \neq 3$ $G(x) = \dfrac{(7 - x)(x - 3)}{x - 3}$

8. $F(x) = x + 5,\ x \neq 4$ $G(x) = \dfrac{(x - 4)(x + 5)}{x - 4}$

9. $F(x) = x,\ x \neq 0$ $G(x) = \dfrac{x^2}{x}$

Graph the following equations.

10. $y^2 = x + 1$
11. $y^2 = x - 1$
12. $x^2 + y^2 = 100$
13. $(x - 6)^2 + y^2 = 100$
14. $x^2 + (y - 6)^2 = 100$
15. $x^2 + 4y^2 = 100$

16. Which of the following are graphs of functions?

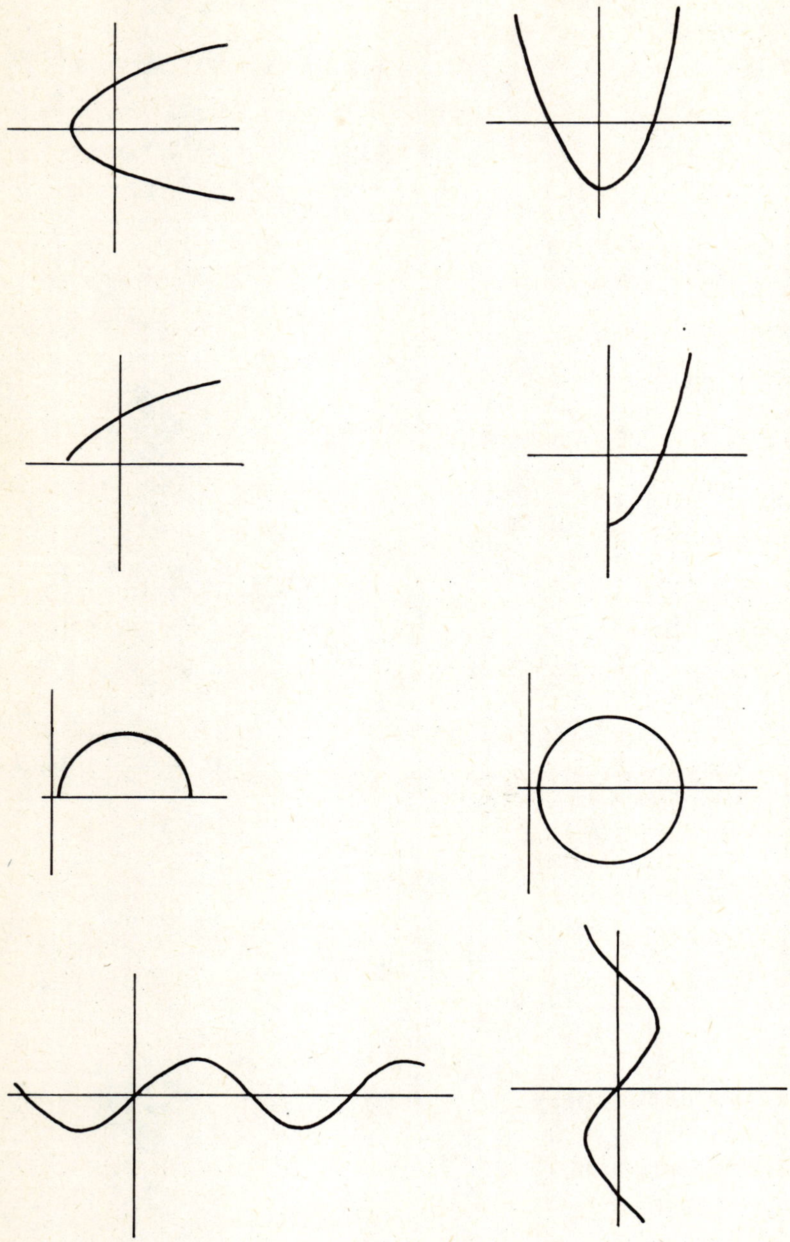

SUMMARY OF LESSON 6

Sometimes it is convenient to restrict the domain of a rule to inputs that make sense in the situation at hand. There is a conventional notation for this [page 95]. This is often called a "restricted domain," while the kind of domain we studied in lesson 5 is called a "natural domain" [page 96].

To graph a rule with a restricted domain, we use only the allowed inputs [page 98]. There are some conventions about how to draw the end points of such curves [pages 98-99]. These are summarized in detail on page 101.

Certain rules use different formulas for different values of the input. We graph these simply by graphing each part separately on the same picture, using the end point conventions mentioned above [page 102]. These rules may have sharp-pointed graphs [page 104].

So far in this module, we have used the words "rule" and "function" interchangeably. Example 18 [page 104] shows the difference. A function is a rule that gives at most one output for each input [page 105]. We can tell whether or not a graph represents a function by applying the vertical line test [page 106].

To graph an equation, we first use algebra to get y alone on the left-hand side [page 108]. In doing this, we may have to split the equation into more than one formula [page 109]. Then we graph it the same way we graphed several-part rules above. This is summarized on page 109. Often, the graphs of equations do not pass the vertical line test.

If you have trouble with any of the exercises, and if the answers don't provide enough help, use the following as a guide for review.

exercises	review
1-3	examples 7-12, pages 99-101
4-6	examples 13-17, pages 102-104
7-9	example 12, page 100
10-15	examples 22-23, pages 108-110
16	examples 19-20, pages 106-107

Lesson 7

GRAPH REASONING II

POSITIVE AND NEGATIVE OUTPUTS

EXAMPLE 1 What values of x make x + 4 positive?

If we put in -4 for x, then x + 4 comes out zero. If we put in any number more negative than -4, then x + 4 comes out negative. Only numbers greater than -4 will make x + 4 come out positive. Answer: $\boxed{x > -4}$.

Notice that this was the same reasoning that we used in lesson 4 to find the natural domain of $\frac{1}{\sqrt{x+4}}$.

EXAMPLE 2 rule F: F(x) = x + 4

What inputs to this rule give positive outputs?

This is really the same question that we asked in example 1, and the answer is the same. But let's see how to answer this question by looking at the graph of rule F.

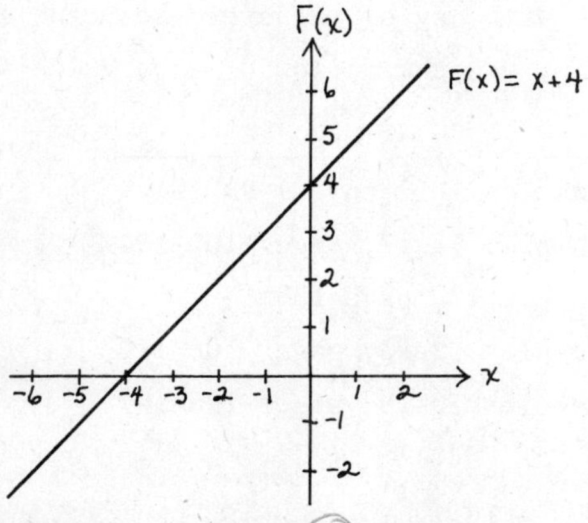

On any graph, the positive outputs are the ones above zero, not including zero (remember that zero is neither positive nor negative).

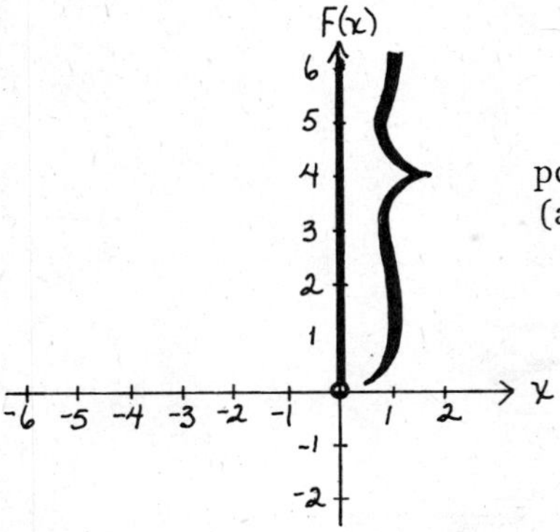

positive outputs
(above the x-axis)

So the (input,output) pairs that have positive outputs are the ones that lie <u>above the x-axis</u>:

These (input,output) pairs have positive outputs.

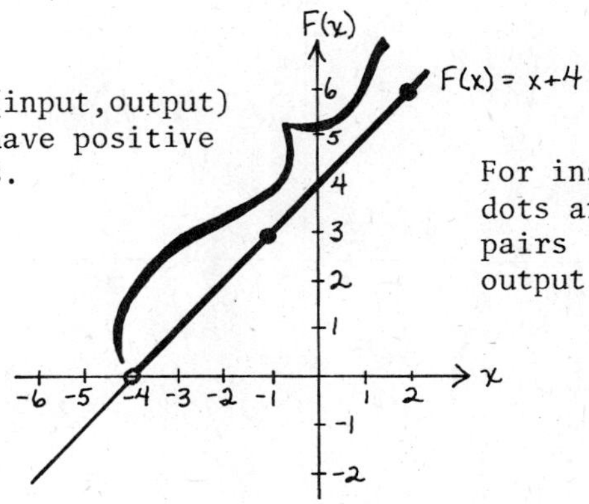

For instance, the big, solid dots are the (input,output) pairs (-1,3), (2,6). Each output is positive.

The corresponding inputs are the ones that lie directly below these pairs on the x-axis:

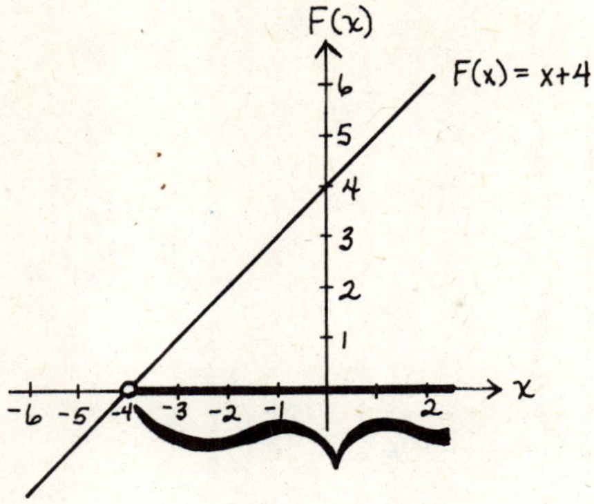

These inputs to rule F give positive outputs. Answer: $\boxed{x > -4}$.

<u>EXAMPLE 3</u> Consider the function graphed below:

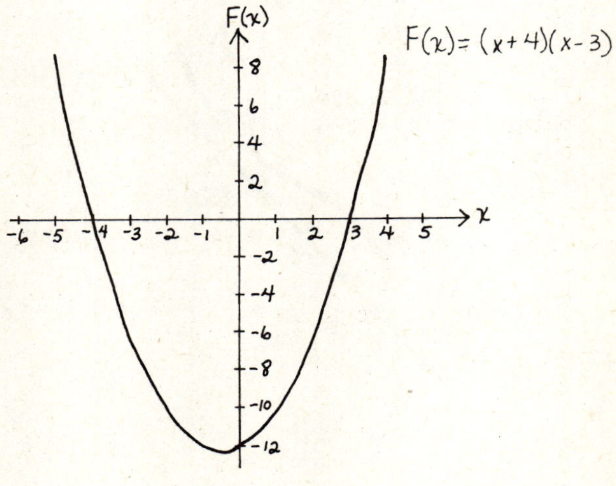

Which inputs to this function give positive outputs?

The (input,output) pairs with positive outputs are the ones that lie above the x-axis:

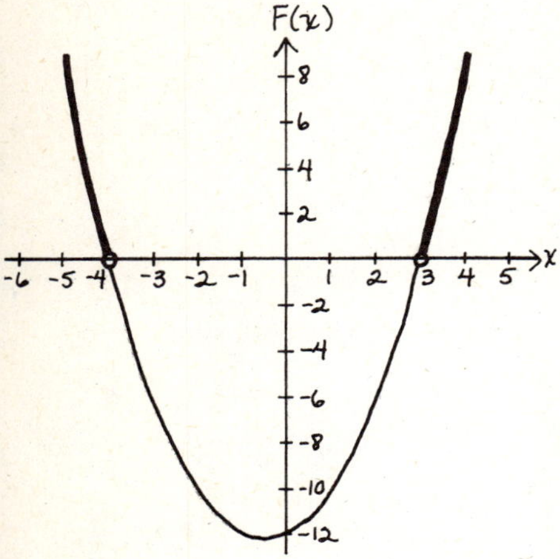

The corresponding inputs are shaded below:

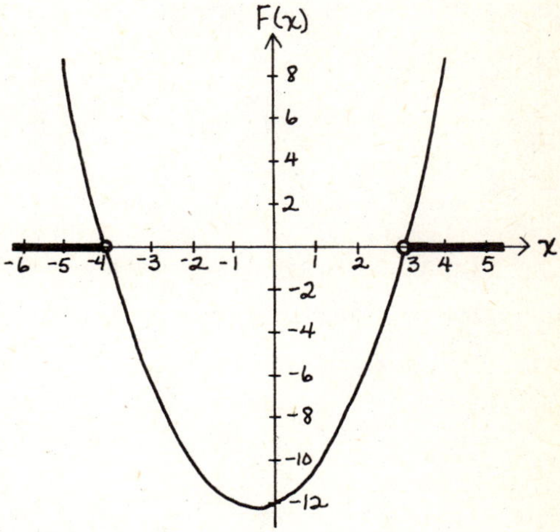

These are the inputs that give positive outputs. Answer: $\boxed{x < -4 \text{ and } x > 3}$.

To check that this kind of "graph reasoning" is correct, let's take some inputs in the range $x < -4$ and $x > 3$, and see if the outputs of rule F are positive.

$$F(-6) = 18, \ F(-5) = 8 \quad F(4) = 8, \ F(5) = 18$$

The output is positive in each case. Also, let's check some inputs that are <u>not</u> in the range $x < -4$ and $x > 3$, and see if the outputs come out negative.

$$F(-3) = -6, \ F(0) = -12, \ F(2) = -6$$

It works.

<u>EXAMPLE 4</u> rule F: $F(x) = (x + 4)(x - 3)$. Which inputs to this function give positive outputs?

The only thing new about this example is the way the question is asked. The question doesn't mention the idea of looking at a graph, so you have to remember this technique for yourself. This is the same function as in example 3, and the graph and the answer are the same. Answer: $\boxed{x < -4, \text{ and } x > 3}$.

EXAMPLE 5 rule F: F(x) = (x + 4)(x - 3). Which inputs to this function give negative outputs?

The reasoning is the same as before, except that negative outputs lie below the x-axis.

The (input,output) pairs with negative outputs are the ones that lie below the x-axis:

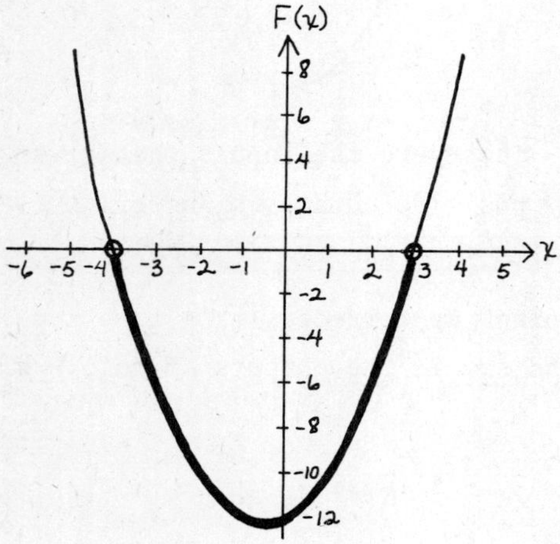

The corresponding inputs are shaded below:

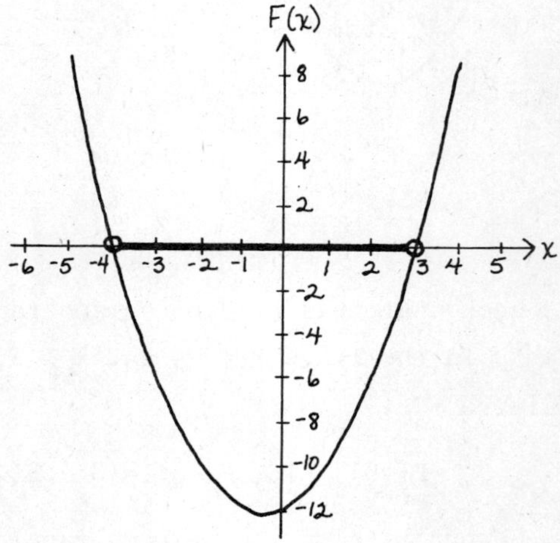

These are the inputs that give negative outputs. Answer: $\boxed{-4 < x < 3}$.

EXAMPLE 6 rule F: F(x) = x(x + 1)(x - 1). Which inputs to this rule give an output of zero? Which inputs give positive outputs? Which inputs give negative outputs?

The first question is simply asking for the zeros of F. We can answer all three questions by sketching the graph of F. (In this case, we can also tell what the zeros of F are by looking at the formula; they are 0, -1, and 1.)

The (input,output) pairs with an output of 0 are indicated below.

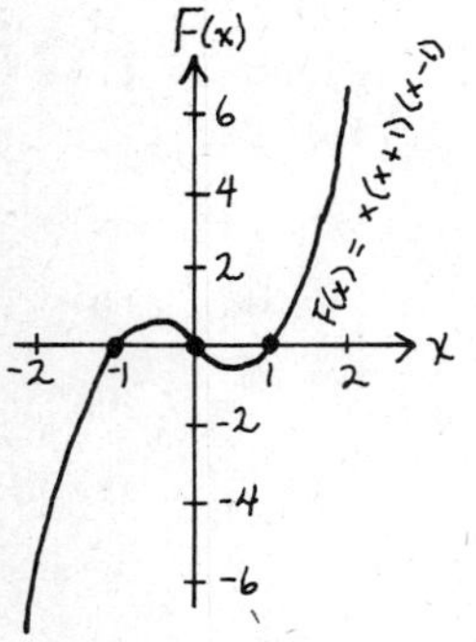

The inputs are:

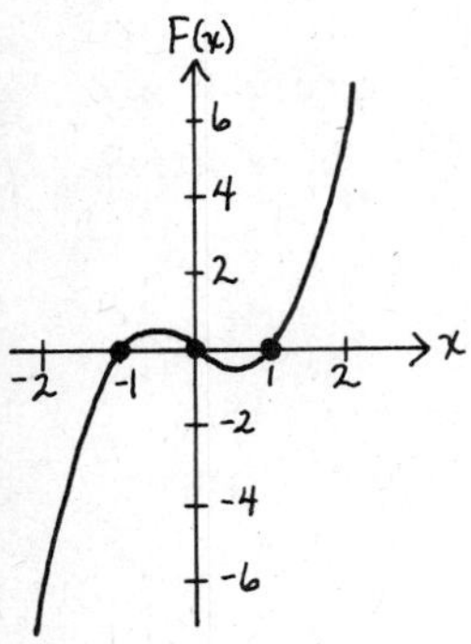

x = -1, 0,1

These are the inputs that give an output of 0.

-when crosses x-axis

The (input,output) pairs with positive outputs are shaded below.

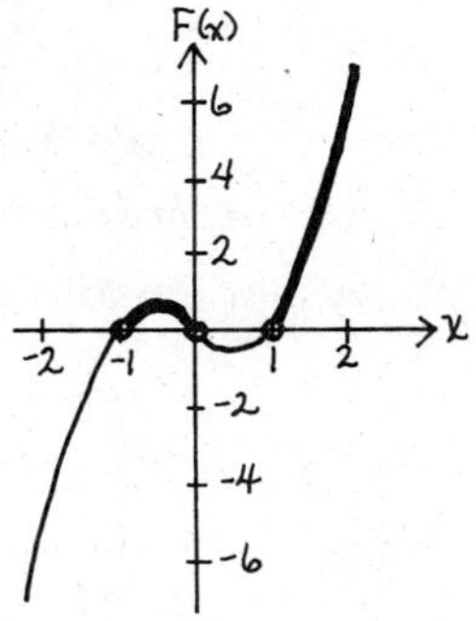

The inputs are:

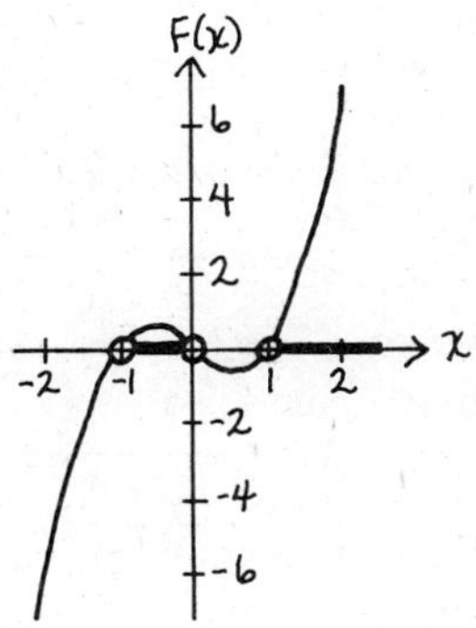

$-1 < x < 0$, $x > 1$

These are the inputs that give positive outputs.

The (input,output) pairs with negative outputs are shaded below.

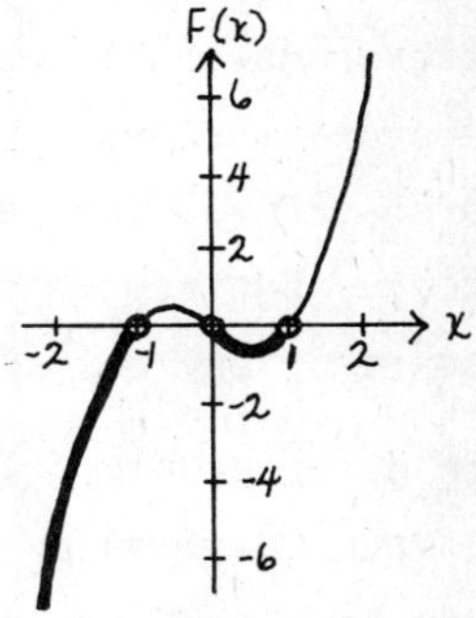

The inputs are:

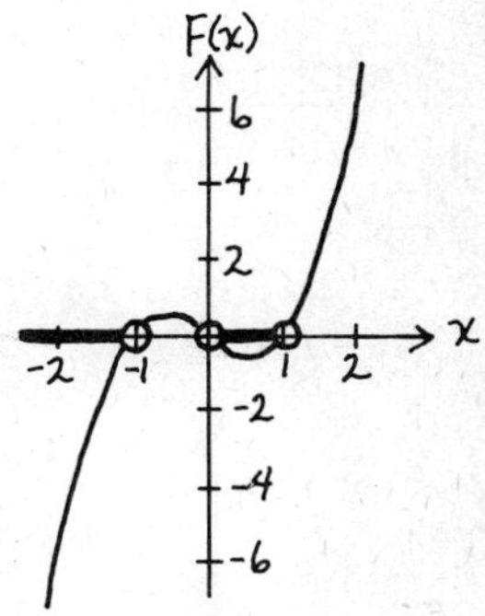

$x < -1$, $0 < x < 1$

These are the inputs that give negative outputs.

Notice that the zeros of F separate the inputs where F(x) is positive from the inputs where F(x) is negative.

zeros of F:

inputs where F(x) is positive:

inputs where F(x) is negative:

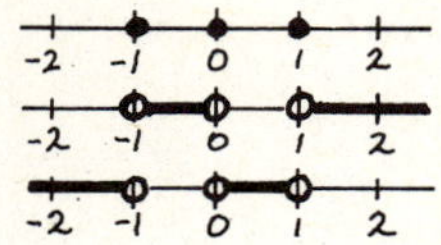

In lesson 3, we saw that zeros often occur between whole numbers. But in this lesson, we are concerned with the basic idea of looking at a graph and seeing where the outputs are positive or negative, not with finding these answers in difficult cases. So in the exercises at the end of this lesson, whenever you are asked to find the inputs where F(x) is positive or negative, the zeros that separate these inputs will be at whole numbers. But remember that this may not be the case in more advanced math texts or in real situations.

In doing problems like this for yourself, you don't need to draw the graph of the function over and over again, as we did in the last example. Once you know what to look for, you can look at a single copy of a graph and tell which inputs give positive, zero, or negative outputs.

<u>EXAMPLE 7</u> rule F: $F(x) = -x(x + 1)(x - 2)$. Which inputs to rule F give positive outputs? What are the zeros of F? Which inputs give negative outputs?

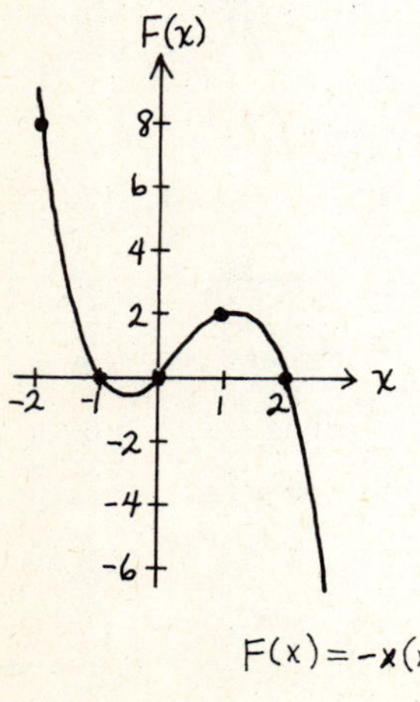

$F(x) = -x(x+1)(x-2)$

The inputs that give positive outputs are $\boxed{x < -1 \text{ and } 0 < x < 2}$. (These are the inputs where the graph lies above the x-axis.)

The zeros of F are $\boxed{x = -1,\ 0,\ 2}$.

The inputs that give negative outputs are $\boxed{-1 < x < 0 \text{ and } x > 2}$. (These are the inputs where the graph lies below the x-axis.)

There is some mathematical terminology that is often used here:

> A group of numbers that lie together on the x-axis is often called an interval. $x \geq -4$ is an interval, for example, as is $0 < x < 2$.
>
> Mathematicians have several ways to say "the inputs to rule F that give positive outputs are $x < -1$ and $0 < x < 2$." One way is to say "the intervals where F is positive are $x < -1$ and $0 < x < 2$." Another way is to say "F is positive on $x < -1$ and $0 < x < 2$." The next example gives some other ways.

(Actually, if you wanted to be technical about it, you could argue that it doesn't make sense to say "F is positive." F is a rule, not a number, and only numbers can be positive or negative. But nevertheless this is what mathematicians say; they aren't always careful with their language. You should get used to it.)

<u>EXAMPLE 8</u> In each case below, give the zeros of F, give the intervals where F is positive, and give the intervals where F is negative.

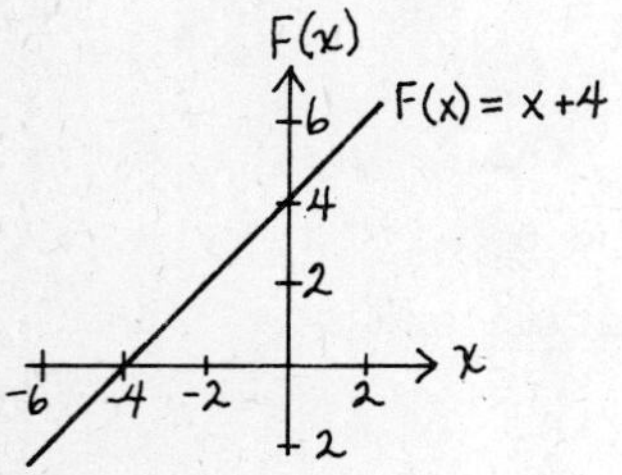

-4 is the only zero of F.

F is positive on $x > -4$.

F is negative on $x < -4$.

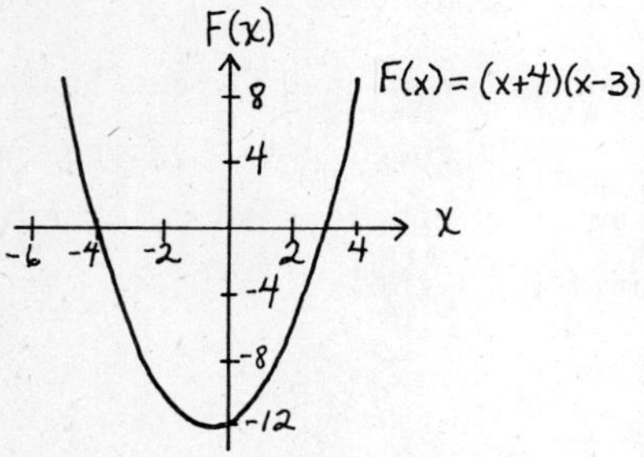

The zeros of F are at $x = -4,\ 3$.

F is positive on the intervals $x < -4$ and $x > 3$.

F is negative on the interval $-4 < x < 3$.

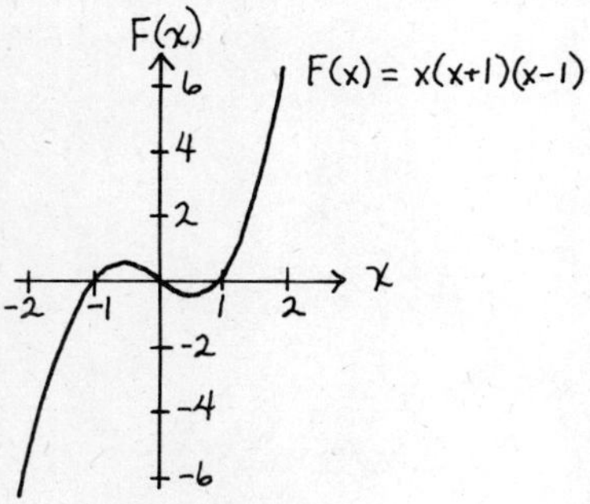

$F(-1) = 0$, $F(0) = 0$, $F(1) = 0$.

F is positive for $-1 < x < 0$ and $x > 1$.

F is negative for $x < -1$ and $0 < x < 1$.

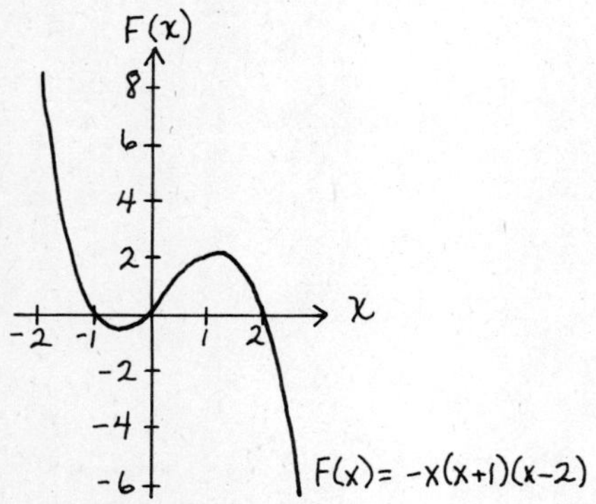

F is zero when $x = -1$, 0, or 2.

F is positive when $x < -1$ and when $0 < x < 2$.

F is negative when $-1 < x < 0$ and and $x > 2$.

Make sure that you understand the language used in example 8. Each answer uses slightly different wording to convey the same kind of information. Which wording to use in any particular case is entirely a matter of personal preference.

Here is a use for the technique of looking at a graph to see when something is positive.

EXAMPLE 9 What is the natural domain of $G(x) = \sqrt{x + 4}$?
We have seen this problem many times, but now we have a new technique for answering it. Instead of reasoning algebraically or looking at the graph of $\sqrt{x + 4}$, we can look at the graph of x + 4. We want the expression under the square root sign to be zero or positive, so we will look at the graph of x + 4 and see when it is zero or positive.

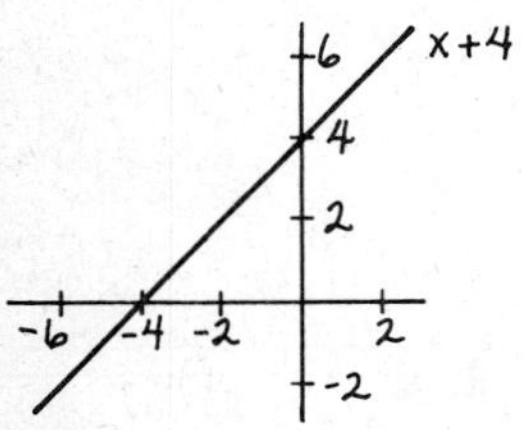

x + 4 is zero when x = -4.
x + 4 is positive when x > -4.

So x + 4 is zero or positive when $x \geqslant -4$.

As we can see from the graph, x + 4 is zero or positive when $x \geqslant -4$. So the natural domain of $\sqrt{x + 4}$ is $\boxed{x \geqslant -4}$.

EXAMPLE 10 What is the natural domain of $G(x) = \sqrt{(x + 4)(x - 3)}$? To answer this, we sketch a graph of (x + 4)(x - 3), and see where it is zero or positive.

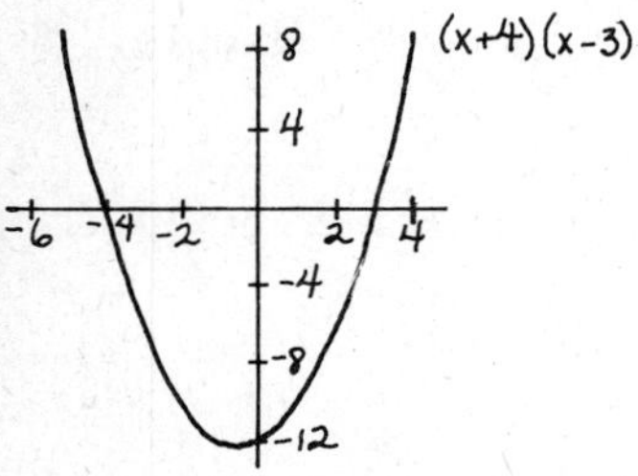

(x + 4)(x - 3) is zero or positive for $x \leqslant -4$ and $x \geqslant 3$.

The expression under the square root sign is zero or positive for $\boxed{x \leqslant -4 \text{ and } x \geqslant 3}$, so this is the natural domain of $G(x) = \sqrt{(x + 4)(x - 3)}$.

<u>EXAMPLE 11</u> What is the natural domain of $G(x) = \dfrac{1}{\sqrt{x(x + 1)(x - 1)}}$?

This time, we can't allow $x(x + 1)(x - 1)$ to be zero, since this would give us a zero in the denominator. Also, we want it to be positive because it is under a square root sign.

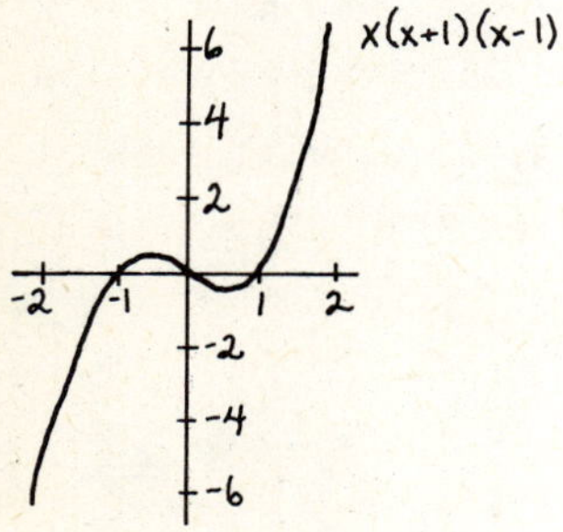

$x(x + 1)(x - 1)$ is positive for $-1 < x < 0$ and $x > 1$.

So $\boxed{-1 < x < 0 \text{ and } x > 1}$ is the natural domain of $G(x) = \dfrac{1}{\sqrt{x(x + 1)(x - 1)}}$.

<u>EXAMPLE 12</u> What is the natural domain of $G(x) = \dfrac{(x + 4)(x - 3)}{\sqrt{x(x + 1)(x - 1)}}$?

The thing to remember here is that we only have to watch out for zeros in the denominator and negatives under the square root sign. There is nothing that can go wrong with the $(x + 4)(x - 3)$ here, since it is not in a denominator or under a square root sign. It doesn't matter whether it is positive, negative, or zero. To find the domain in this case, we only have to look at the $x(x + 1)(x - 1)$, which we did in the last example. So the answer is the same, $\boxed{-1 < x < 0 \text{ and } x > 1}$.

INTERVALS OF INCREASE AND DECREASE

Here is another kind of information we can get by looking at a graph:

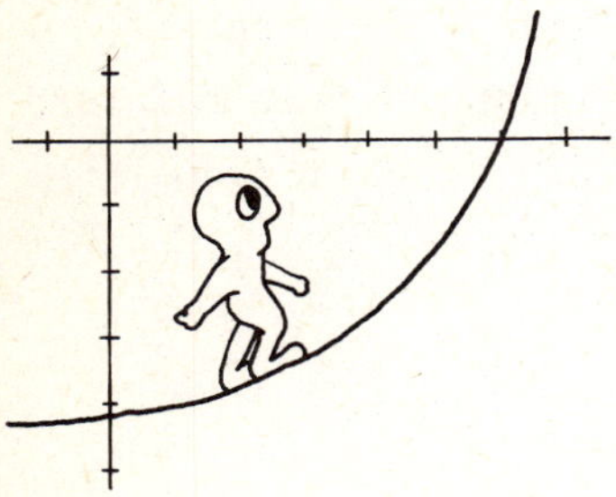

We say that a function is increasing from left to right (or simply increasing) if its graph slopes up as we move along it from left to right.

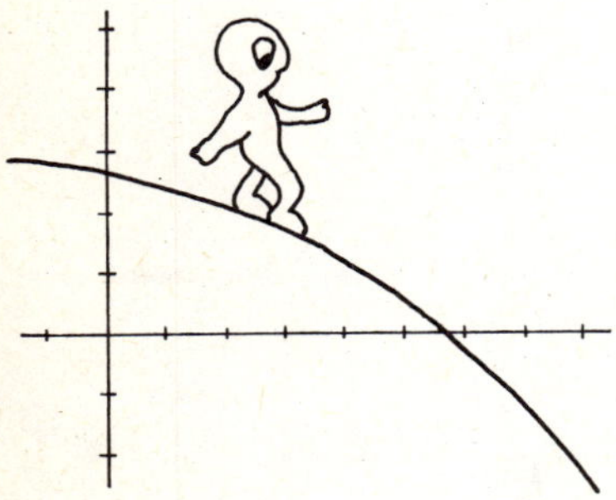

We say that a function is decreasing from left to right (or simply decreasing) if its graph slopes down as we move along it from left to right.

(We could also say that a graph is increasing if its slope is positive, and it is decreasing if its slope is negative. We already know how to find the slope of a straight line, and if you study calculus, you will learn how to find slopes at different points on a curve. But for this module, we will simply look at the graph, as described above.)

EXAMPLE 13 rule F: $F(x) = x^3 + x$. Is this function increasing?

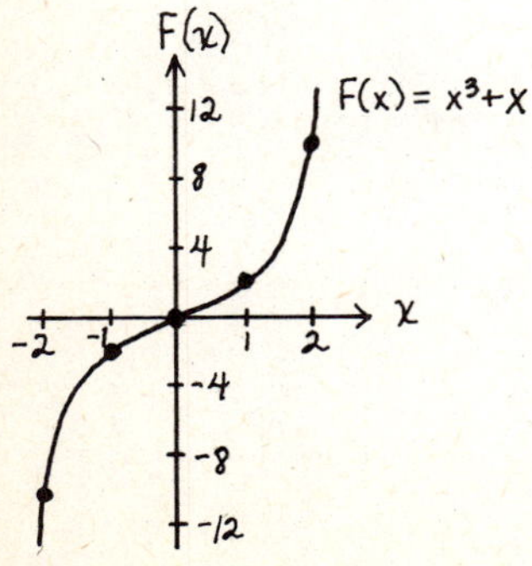

The graph of F slopes upwards from left to right, so F is increasing. Answer: yes.

EXAMPLE 14 rule F: $F(x) = x^2$. Is this function increasing?

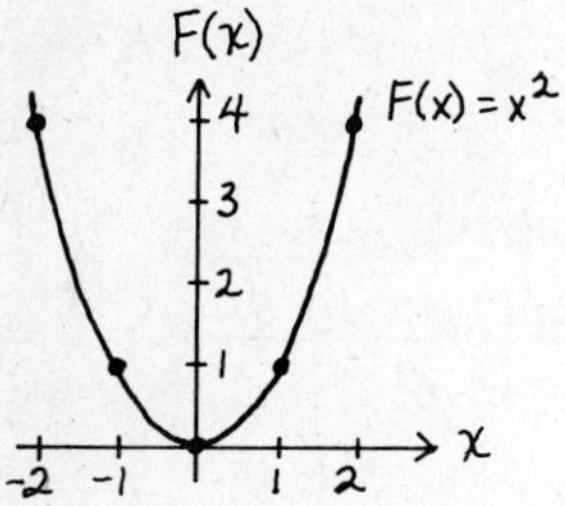

To the right of the F(x)-axis, the graph is sloping up from left to right. But to the left of the F(x)-axis, it is sloping down from left to right. Answer: no.

But part of the curve is increasing.

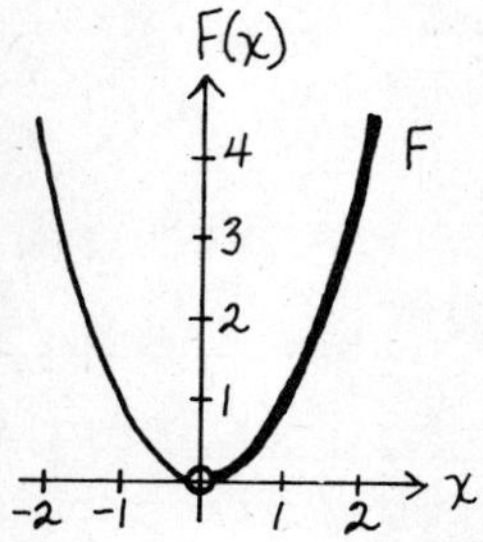

The corresponding inputs are the interval $x > 0$.

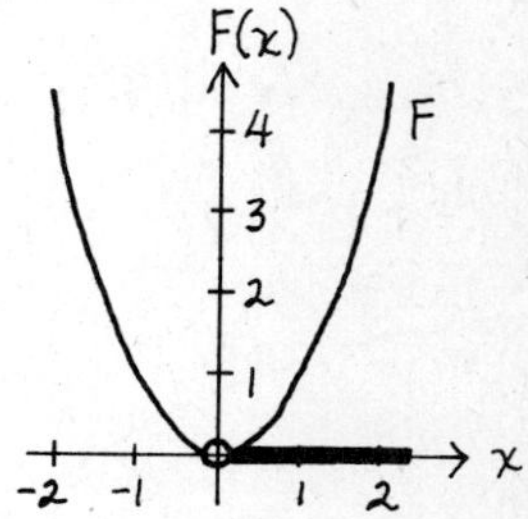

In this case, we say that F is increasing on the interval $x > 0$, or simply F is increasing on $x > 0$.

Also, part of the curve is decreasing.

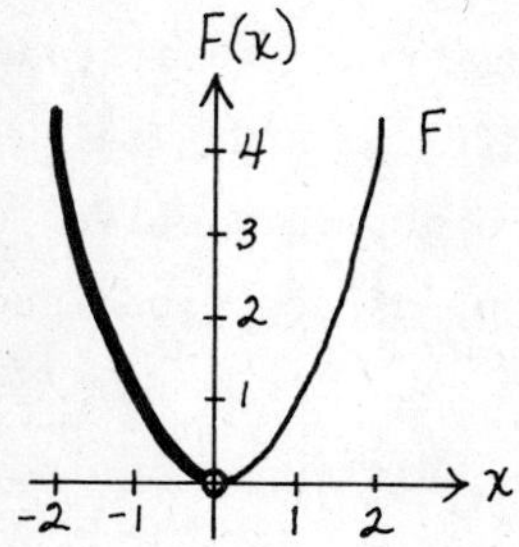

The corresponding inputs are the interval $x < 0$.

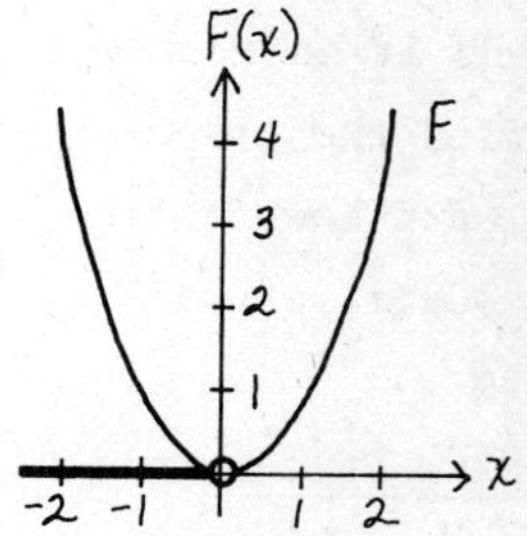

So we say that F is decreasing on the interval $x < 0$, or simply F is decreasing on $x < 0$.

Notice that right at the bottom point on the curve, the function is neither increasing or decreasing.

EXAMPLE 15 Find the intervals where $F(x) = x^3 - 3x$ is increasing and where it is decreasing.

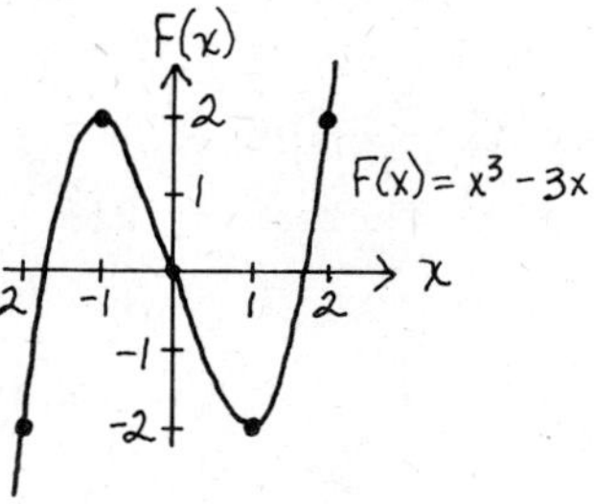

These are the parts of the curve where F is increasing.

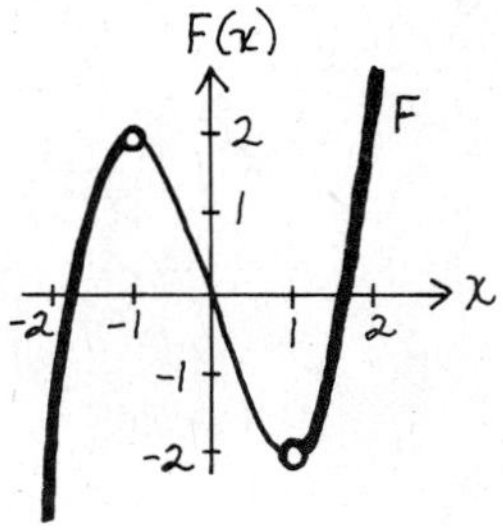

So F is increasing from left to right on $\boxed{x < -1}$ and $\boxed{x > 1}$.

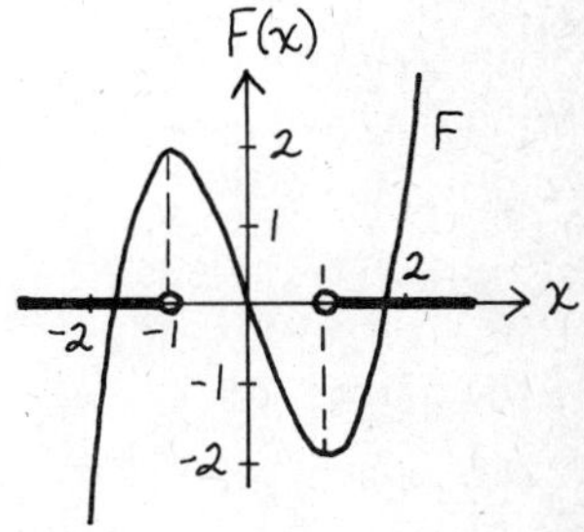

This is the part of the curve where F is decreasing.

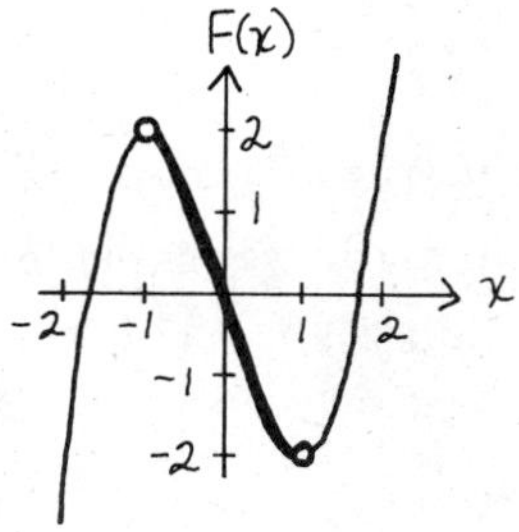

So F is decreasing from left to right on $\boxed{-1 < x < 1}$.

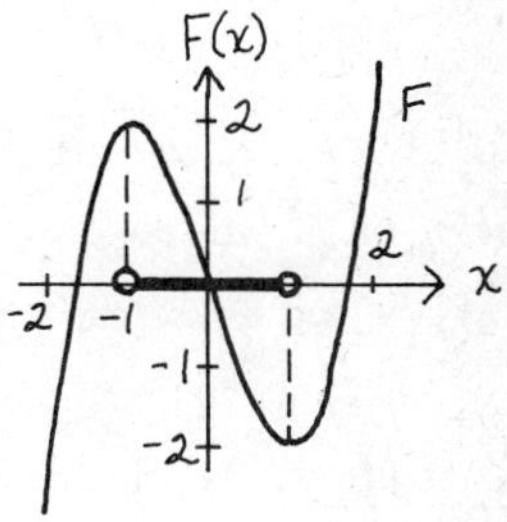

The peak points  and the valley points 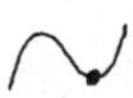 on the graph separate the increasing parts of the curve from the decreasing parts. (On the graph above, these are not the highest or lowest points.) On the graph of F above, can we be sure that the inputs at these points are exactly -1 and 1? We can't, although we could check the graph close to these points to be more certain. In calculus, you will learn how to find these inputs exactly. For now, we are concerned with learning the basic principles of interpreting graphs, not with using graphs to find exact answers. In the exercises at the end of this lesson, whenever you are asked to find the intervals where a function is increasing or decreasing, the inputs that separate these intervals will be at whole numbers. You will learn how to handle more difficult cases in calculus.

Notice that the peak points and the valley points of F are not the same as the zeros, and the intervals where F is increasing are not the same as the intervals where F is positive.

The parts of the curve where F is increasing:

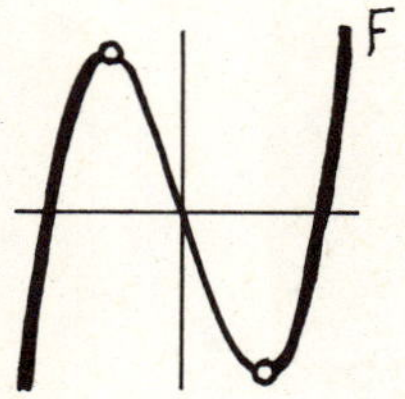

→ The intervals where F is increasing:

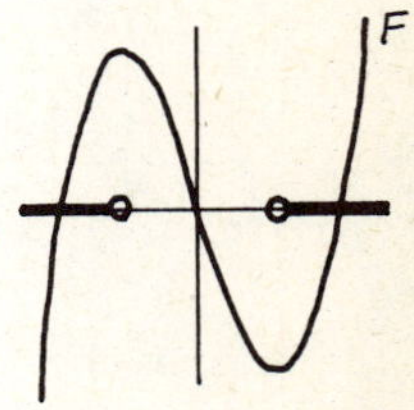

The parts of the curve where F is positive:

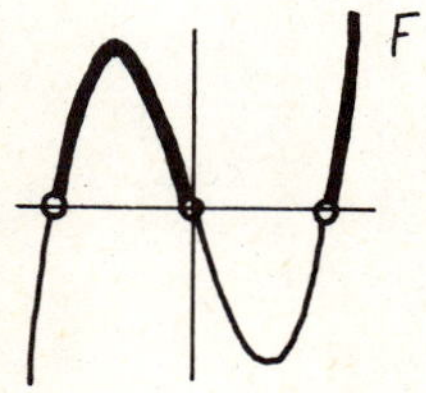

→ The intervals where F is positive:

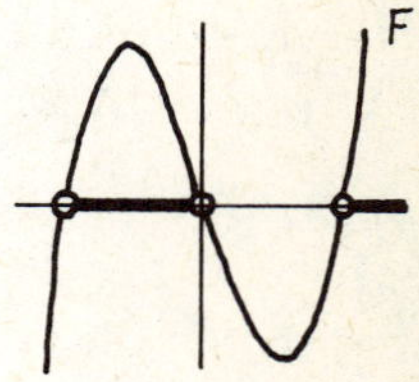

Similarly, the intervals where F is decreasing are not the same as the intervals where F is negative. (Check this for yourself on the graph of $F(x) = x^3 - 3x$ at the beginning of example 15.)

<u>EXAMPLE 16</u> Find the zeros of $F(x) = x^2(x - 3)$, and find the intervals where F is positive and negative. Then find the peak and valley points, and find the intervals where F is increasing and decreasing.

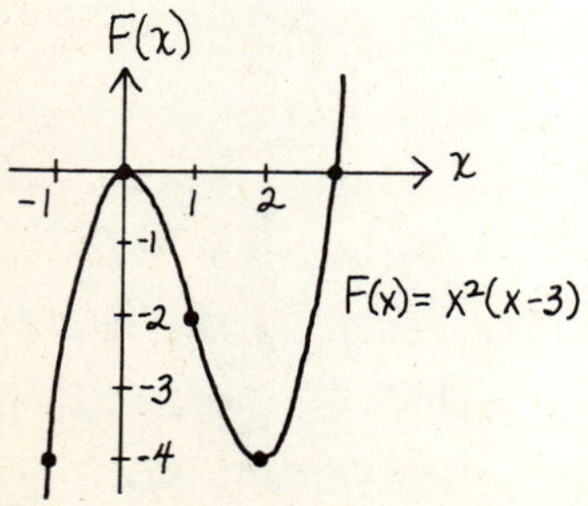

The zeros of F are 0 and 3.
F is positive on $x > 3$.
F is negative on $x < 0$ and $0 < x < 3$.

The peak and valley points are (0,0) and (2,-4).
F is increasing on $x < 0$ and $x > 2$.
F is decreasing on $0 < x < 2$.

You might have been tempted to say that F is negative on $x < 3$. But $x < 3$ includes the input 0, and the output there is not negative. (It's not positive either—it's zero.)

EXERCISES (Answers begin on page 137.)

Find the zeros of each function, and give the intervals where each function is positive and the intervals where it is negative.

1. $F(x) = 8 - x$
2. $F(x) = \frac{1}{2}x - 1$
3. $F(x) = (x + 1)(x - 1)$
4. $F(x) = x(2 - x)$
5. $F(x) = x(x - 1)(x - 2)$
6. $F(x) = -x(x - 2)(x - 3)$

Give the natural domain of each of the following functions.

7. $F(x) = \sqrt{8 - x}$
8. $G(x) = \dfrac{1}{\sqrt{8 - x}}$
9. $F(x) = \sqrt{(x + 1)(x - 1)}$
10. $G(x) = \dfrac{1}{\sqrt{(x + 1)(x - 1)}}$
11. $F(x) = \sqrt{x(x - 1)(x - 2)}$
12. $G(x) = \dfrac{1}{\sqrt{x(x - 1)(x - 2)}}$

In each case, find the peak and valley points, and give the intervals where the function is increasing and where it is decreasing.

13. $F(x) = (x + 1)(x - 1)$
14. $F(x) = x(2 - x)$
15. $F(x) = x^2(2x - 3)$
16. $F(x) = 9x + 3x^2 - x^3$
17. $F(x) = 8 - x$
18. $F(x) = \frac{1}{2}x - 1$

In each case, find the zeros of F, and find the intervals where F is positive and negative. Then find the peak and valley points, and give the intervals where F is increasing and decreasing.

19. $F(x) = (x + 1)(x - 2)^2$
20. $F(x) = -x^2(x + 3)$

Save graphs of these exercises, and turn to G.S. page 95.

ANSWERS TO EXERCISES

(1) $F(x) = 8 - x$

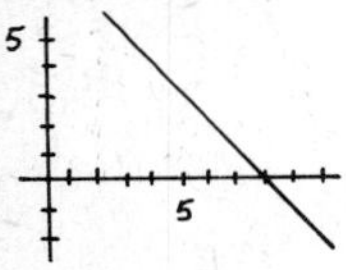

Zeros: $x = 8$
positive: $x < 8$
negative: $x > 8$

(2) $F(x) = \frac{1}{2}x - 1$

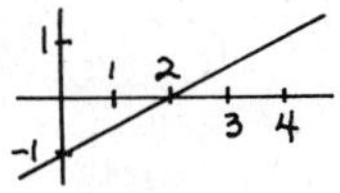

zeros: $x = 2$
positive: $x > 2$
negative: $x < 2$

(3) $F(x) = (x+1)(x-1)$

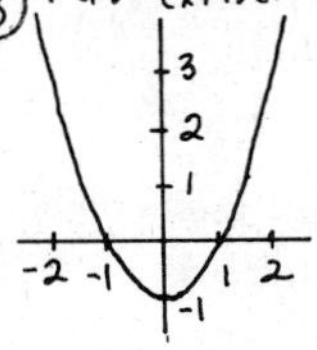

zeros: $-1, 1$
positive: $x < -1, x > 1$
negative: $-1 < x < 1$

(4) $F(x) = x(2-x)$

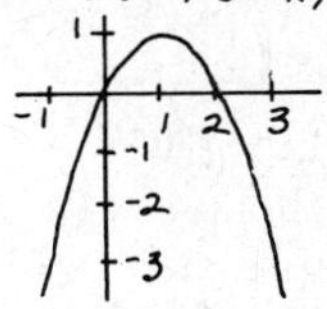

zeros: 0, 2
positive: $0 < x < 2$
negative: $x < 0, x > 2$

(5) $F(x) = x(x-1)(x-2)$

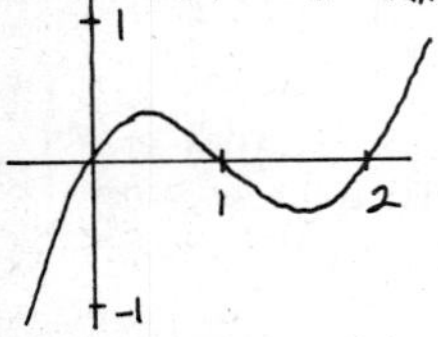

zeros: 0, 1, 2
positive: $0 < x < 1, x > 2$
negative: $x < 0, 1 < x < 2$

(6) $F(x) = -x(x-2)(x-3)$

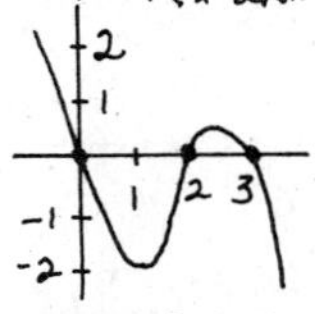

zeros: 0, 2, 3
positive: $x < 0, 2 < x < 3$
negative: $0 < x < 2, x > 3$

(7)

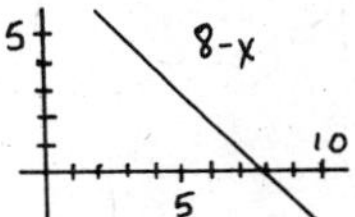

$8 - x$ is positive or zero on $x \le 8$
So the natural domain of $\sqrt{8-x}$
is $x \le 8$.

(8) From the same graph as in #7, we can see that $8 - x$ is positive on $x < 8$. So the natural domain of $\frac{1}{\sqrt{8-x}}$ is $x < 8$.

(9)

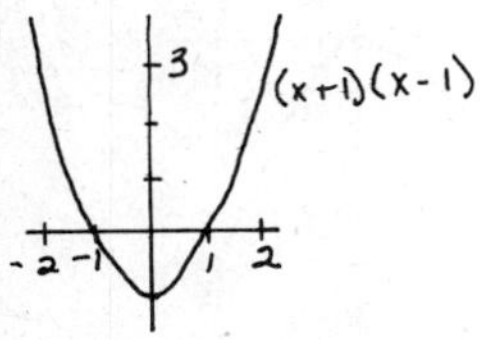

$(x+1)(x-1)$ is positive or zero on $x \le -1$ and $x \ge 1$
So the natural domain of $\sqrt{(x+1)(x-1)}$ is $x \le -1$ and $x \ge 1$.

(10) From the same graph as in #9, we can see that $(x+1)(x-1)$ is positive on $x < -1$ and $x > 1$. So the natural domain of $\frac{1}{\sqrt{(x+1)(x-1)}}$ is $x < -1$ and $x > 1$.

(11)

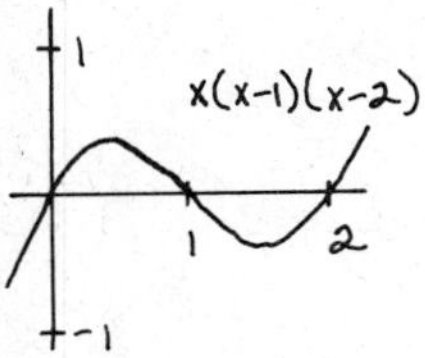

$x(x-1)(x-2)$ is positive or zero on $0 \le x \le 1$ and $x \ge 2$
So the natural domain of $\sqrt{x(x-1)(x-2)}$ is
$0 \le x \le 1$ and $x \ge 2$.

(12) From the same graph as in #11, we can see that $x(x-1)(x-2)$ is positive on $0 < x < 1$ and $x > 2$. So the natural domain of $\frac{1}{\sqrt{x(x-1)(x-2)}}$ is $0 < x < 1$ and $x > 2$.

(13) $F(x) = (x+1)(x-1)$

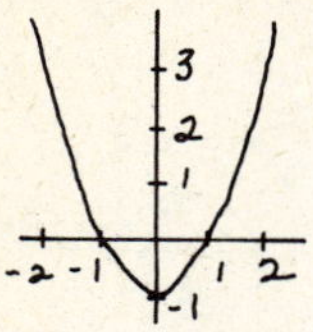

valley point: (0, −1)
increasing: $x > 0$
decreasing: $x < 0$

(14) $F(x) = x(2-x)$

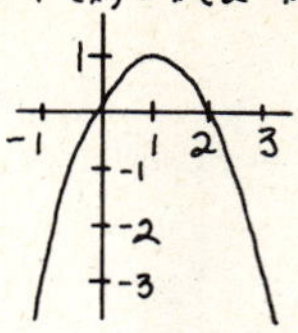

peak point: (1, 1)
increasing: $x < 1$
decreasing: $x > 1$

(15) $F(x) = x^2(2x-3)$

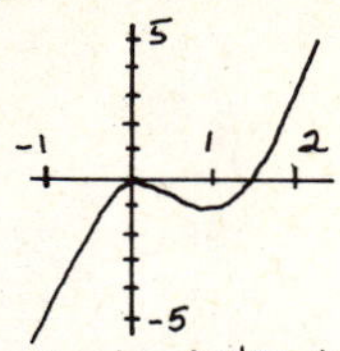

peak point: (0, 0)
valley point: (1, −1)
increasing: $x < 0$ and $x > 1$
decreasing: $0 < x < 1$

(16) $F(x) = 9x + 3x^2 - x^3$

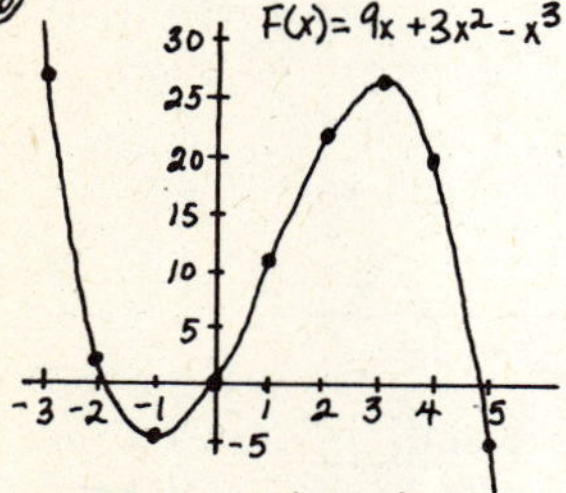

peak point: (3, 27)
valley point: (−1, −5)
increasing: $-1 < x < 3$
decreasing: $x < -1$ and $x > 3$

(17) $F(x) = 8 - x$

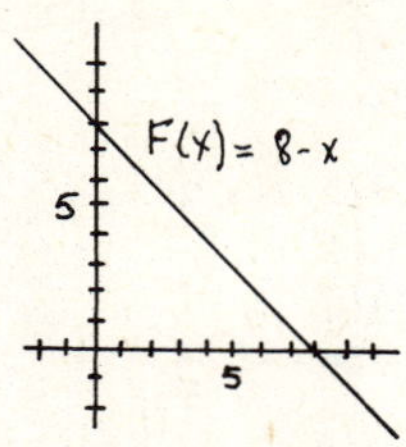

no peak or valley points
increasing: nowhere
decreasing: everywhere

(18) $F(x) = \frac{1}{2}x - 1$

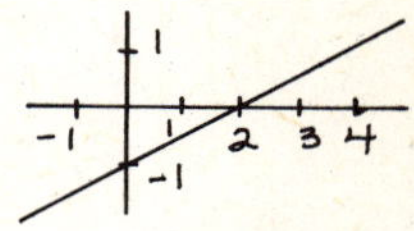

no peak or valley points
increasing: everywhere
decreasing: nowhere

(19) $F(x) = (x+1)(x-2)^2$

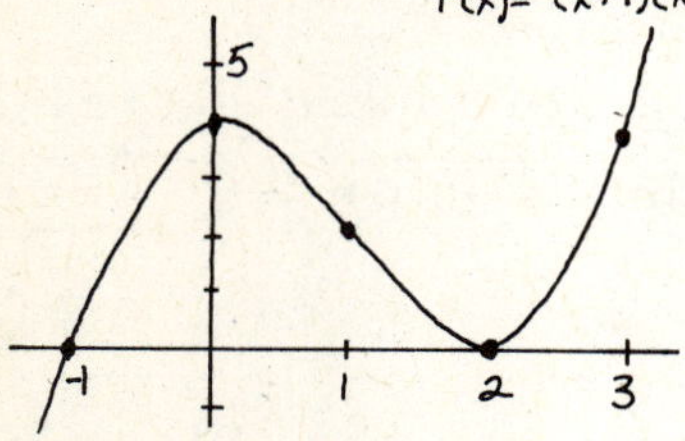

zeros: −1, 2
positive: $-1 < x < 2$, $x > 2$
negative: $x < -1$
peak point: (0, 4)
valley point: (2, 0)
increasing: $x < 0$, $x > 2$
decreasing: $0 < x < 2$

(20) $F(x) = -x^2(x+3)$

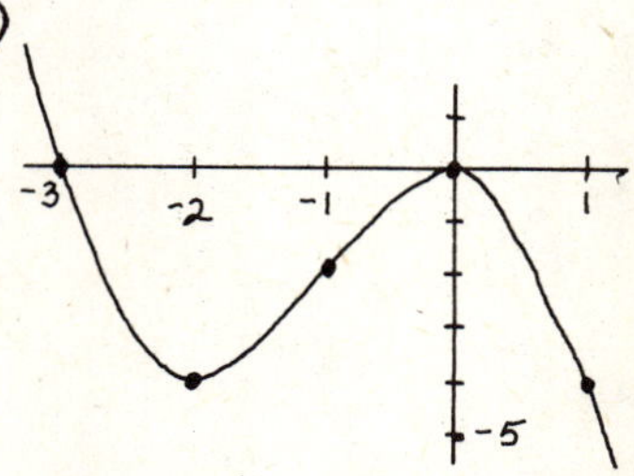

zeros: −3, 0
positive: $x < -3$
negative: $-3 < x < 0$, $x > 0$
peak point: (0, 0)
valley point: (−2, −4)
increasing: $-2 < x < 0$
decreasing: $x < -2$, $x > 0$

ADDITIONAL EXERCISES

Find the zeros of each function, and specify the intervals where each function is positive and the intervals where it is negative.

1. $F(x) = 7 - x$
2. $F(x) = \frac{1}{3}x - 1$
3. $F(x) = x(x + 2)$
4. $F(x) = (1 - x)(x - 3)$
5. $F(x) = (x - 1)(x - 2)(x - 3)$
6. $F(x) = (x + 1)(1 - x)(x - 2)$

Give the natural domain of each of the following functions.

7. $F(x) = \sqrt{7 - x}$
8. $G(x) = \dfrac{1}{\sqrt{7 - x}}$
9. $F(x) = \sqrt{x(x + 2)}$
10. $G(x) = \dfrac{1}{\sqrt{x(x + 2)}}$
11. $F(x) = \sqrt{(x - 1)(x - 2)(x - 3)}$
12. $G(x) = \dfrac{1}{\sqrt{(x - 1)(x - 2)(x - 3)}}$

In each case, find the intervals where F is increasing and the intervals where F is decreasing.

13. $F(x) = x(x - 2)$
14. $F(x) = (1 - x)(x - 3)$
15. $F(x) = (x - 1)^2(2x - 5)$
16. $F(x) = (x + 1)(11 + x - x^2)$
17. $F(x) = 7 - x$
18. $F(x) = \frac{1}{3}x - 1$

In each case, find the zeros of F, and find the intervals where F is positive, negative, increasing, and decreasing.

19. $F(x) = -(x + 1)(x - 2)^2$
20. $F(x) = x^2(x + 3)$

SUMMARY OF LESSON 7

The inputs to a function that give positive outputs are the inputs where the graph of the function lies above the x-axis [page 121]. The inputs that give negative outputs are the ones where the graph lies below the x-axis [page 122]. We specify these groups of inputs with inequality notation.

A group of inputs that lie together on the x-axis is called an "interval" [page 127]. Instead of talking about "the inputs to F that give positive outputs," mathematicians usually say "the intervals where F is positive" [page 127]. There are many slight variations on this language [page 128].

Finding the intervals where a formula is positive can help in determining the natural domain of a function [page 129].

A function is "increasing" if its graph slopes up as we move along it from left to right. A function is "decreasing" if its graph slopes down as we move along it from left to right [page 131]. Many functions are increasing on part of their domain, and decreasing on another part [page 132]. We indicate the inputs where F is increasing (or decreasing) with inequality notation. We talk about the "intervals where F is increasing," and the "intervals where F is decreasing" [page 132].

The peak points 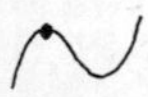 and the valley points 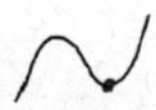 on a graph separate the increasing parts of the curve from the decreasing parts. The inputs at these points separate the intervals where the function is increasing from the intervals where it is decreasing [page 133]. Right at the peak or valley point, the function is neither increasing nor decreasing [page 132]. In this module, the inputs that separate these intervals will always be at whole numbers [page 134]. You will learn how to handle more difficult cases in calculus.

If you have trouble with any of the exercises, and if the answers don't give enough help, use the following as a guide for review.

exercises	review
1-6	pages 39-41; pages 120-128
1-2	example 2, page 120
3-4	example 3, page 122; example 5, page 124
5-6	examples 6-7, pages 124-126
7-12	examples 9-12, pages 129-130
13-20	pages 129-135
13-14	example 14, page 132
15-16	example 15, page 133
19-20	example 16, page 134

Lesson 8

ABSTRACT GRAPHS

This lesson contains the kinds of graphs that often appear in advanced math books (particularly calculus books). These are basically the same as the graphs we have been studying, but sometimes they are more "abstract." That is, some of the information that we are used to seeing on a graph—such as the formula of the function, or the numbers along the axes, or the labels on the axes—may be missing, and we may have to use new methods to answer questions about the graph.

POINT LABELING

EXAMPLE 1 Label the indicated points with their coordinates.

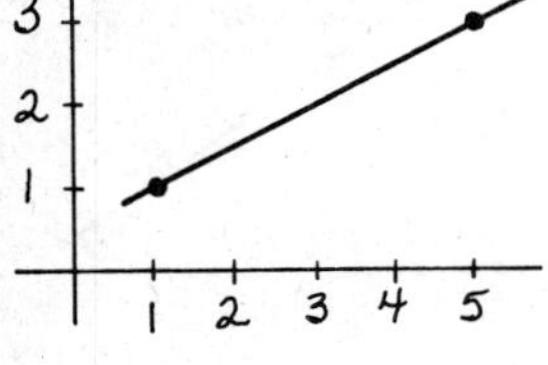

In this case, we can read the inputs and outputs directly from the numbers along the axes.

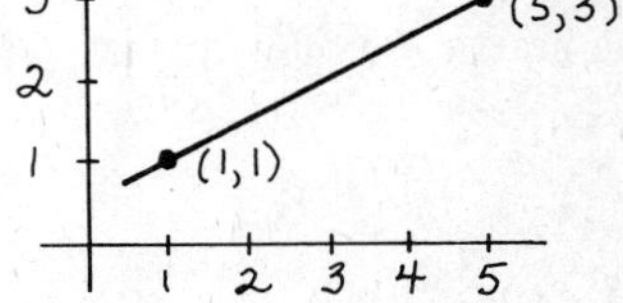

EXAMPLE 2 Label the indicated points with their coordinates.

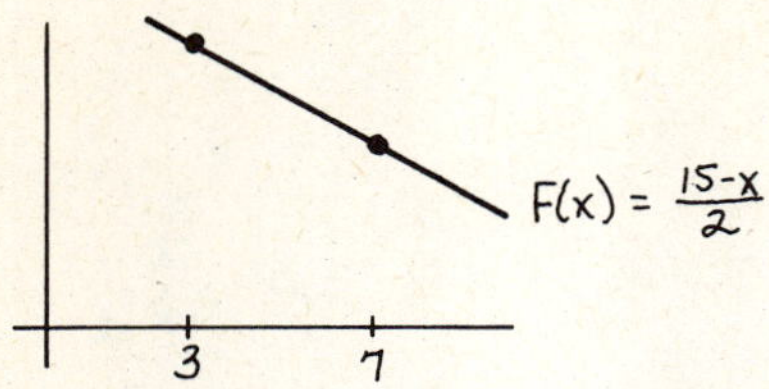

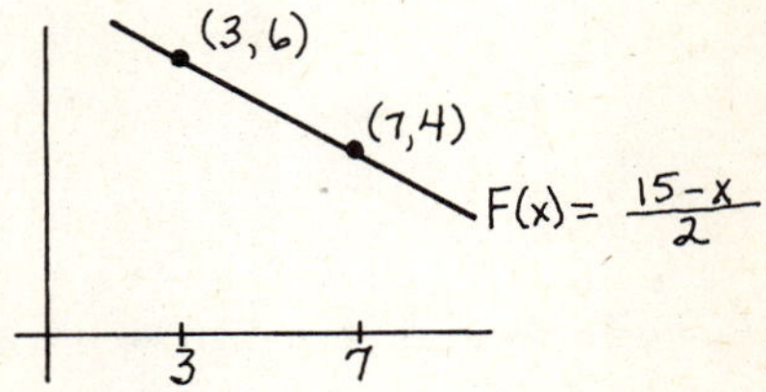

Math books often give graphs without numbers along the F(x)-axis. In this case, we can use the formula to figure the outputs at the given points.
F(3) = 6, so the coordinates of the left-hand point are (3,6).
F(7) = 4, so the coordinates of the right-hand point are (7,4).

EXAMPLE 3 Label the indicated points with their coordinates.

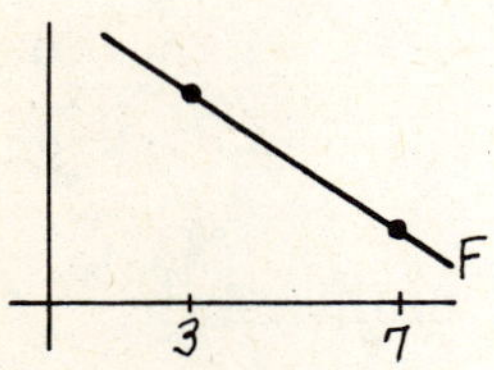

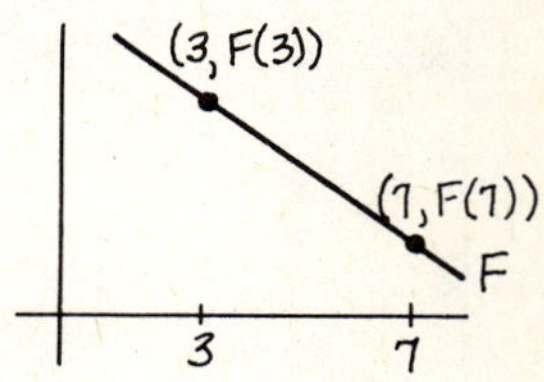

Graphs are often presented with no numbers along the F(x)-axis and with no formula for the function. How can we figure the outputs? In this case, the best answer we can give for the output of F given input 3 is F(3). When the input is 7, the output is F(7). To give numerical answers we would have to know a formula for F, as in the last example.

EXAMPLE 4 Label the indicated points with their coordinates.

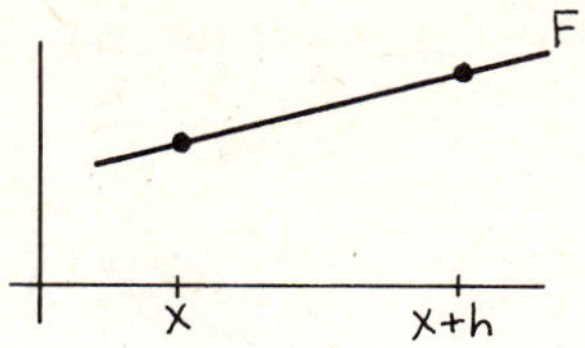

Using letters to stand for inputs on a graph is another common practice. Now, when the input is x, the best answer we can give for the output is F(x). When the input is x + h, the output is F(x + h).

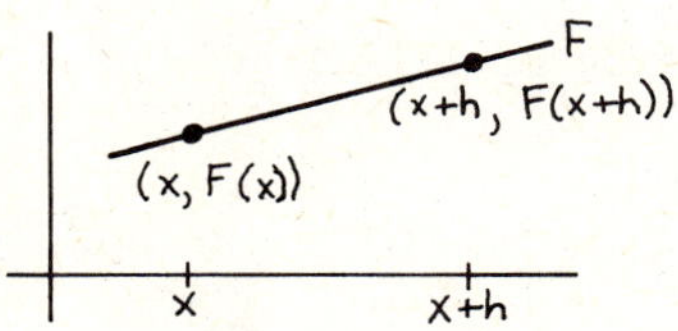

EXAMPLE 5 Label the indicated points with their coordinates.

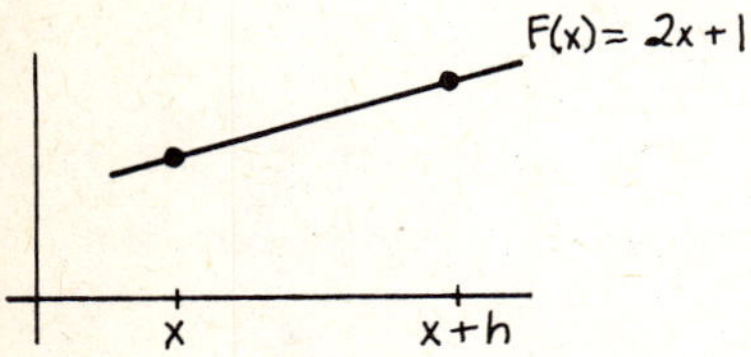

Again we have letters that stand for inputs. But this time we also have a formula for F. So instead of giving simply F(x) and F(x + h) as outputs, we can work out formulas for these outputs.

$F(x) = 2x + 1$

$F(\quad) = 2(\quad) + 1$

$F(x + h) = 2(x + h) + 1 = 2x + 2h + 1$

(The process of re-writing the formula with blanks, and then putting x + h into the blanks, was described in Module X, lesson 7. Review this if you have any questions.)

So when the input is x, the output is 2x + 1.

When the input is x + h, the output is 2x + 2h + 1.

To give numerical answers for these outputs, we would have to know what numbers x and h stand for.

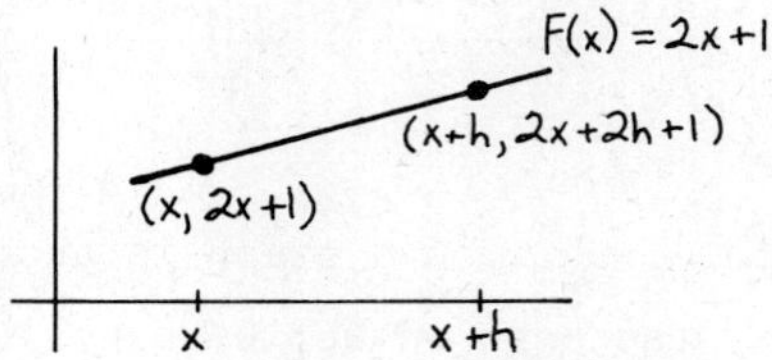

SLOPE

EXAMPLE 6 What is the slope of this line?

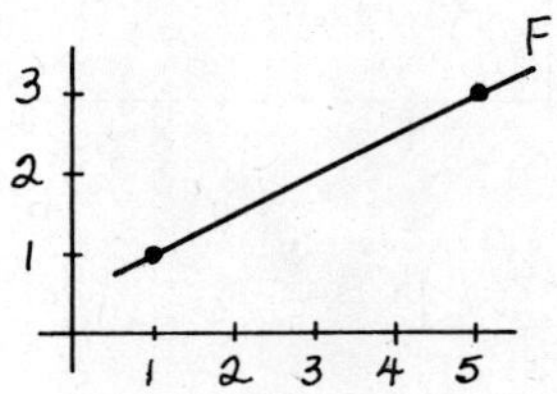

To find the slope of a straight line, we start with two points on the line and find Δy and Δx between them. Δy is the vertical separation between the points, and Δx is the horizontal separation.

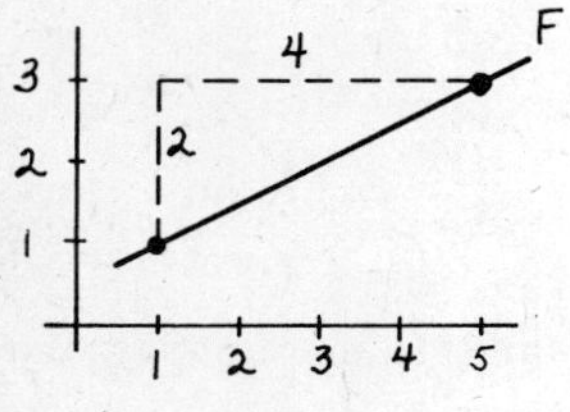

For the points on the line to the left, Δy is 2 and Δx is 4. (If you have any questions about this, review lesson 5 in Module III, Linear Equations and Lines.)

Once we have found Δy and Δx, we can calculate the slope:

$$\text{slope} = \frac{\Delta y}{\Delta x}$$

So in this case, the slope is $\frac{2}{4}$, or $\frac{1}{2}$.

There is another way to find Δy and Δx that will be useful in this lesson: take the coordinates of the right-hand point, and subtract the coordinates of the left-hand point.

	(5,3)
subtract:	(1,1)
	4,2

The difference between the x-coordinates is 4, so Δx is 4. The difference between the y coordinates is 2, so Δy is 2. The slope is $\frac{\Delta y}{\Delta x}$, so the slope is $\frac{2}{4}$, or $\frac{1}{2}$. This is the same answer as before.

What happens if we are trying to use this method, but we do the subtraction in the opposite order?

	(1,1)	Remember: to subtract,	(1, 1)
subtract:	(5,3)	change signs and add.	(-5,-3)
			-4,-2

Our answers for Δy and Δx are the opposite of what they were before. But when we compute $\frac{\Delta y}{\Delta x}$ to get the slope, we get $\frac{-2}{-4}$, which equals $\frac{1}{2}$. This is the same answer as before.

It doesn't really matter which coordinates you write down first. The slope will come out the same either way. In the following examples, and in the exercises at the end of this lesson, we will always start with the right-hand point and subtract the coordinates of the left-hand point. But you could also do it the other way.

EXAMPLE 7 Find the slope of the following line by finding Δy and Δx between the indicated points.

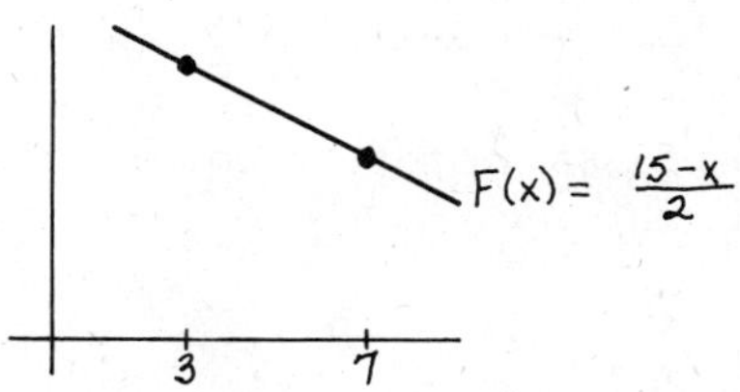

Since we can't count along the F(x)-axis, we will have to use the subtraction method. But first we need the coordinates of the points. We found these in example 2 by using the formula for F.

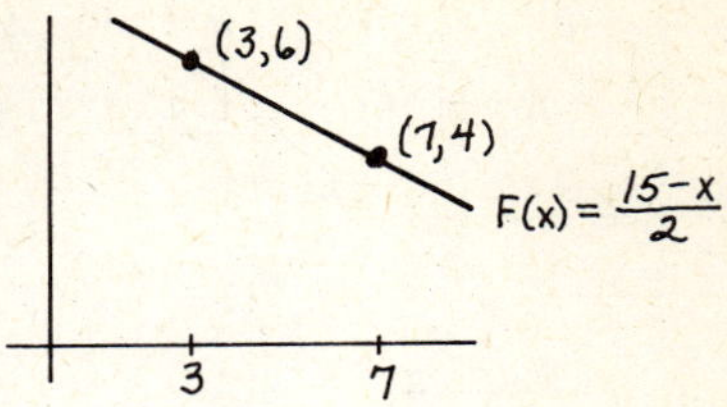

We take the coordinates of the right-hand point, and subtract the coordinates of the left-hand point.

$$\text{Subtract}\ \begin{array}{r}(7,4)\\ \underline{(3,6)}\\ 4,-2\end{array}.\qquad \text{Slope} = \frac{\Delta y}{\Delta x} = \frac{-2}{4} = \boxed{-\frac{1}{2}}\ .$$

(We could also find the slope by putting the formula for F in y = mx + b form. $y = \frac{15 - x}{2}$ is the same as $y = -\frac{1}{2}x + \frac{15}{2}$. The number multiplying the x is the slope. This was covered in Module III, lesson 6.)

<u>EXAMPLE 8</u> Find the slope of the following line by finding Δy and Δx between the indicated points.

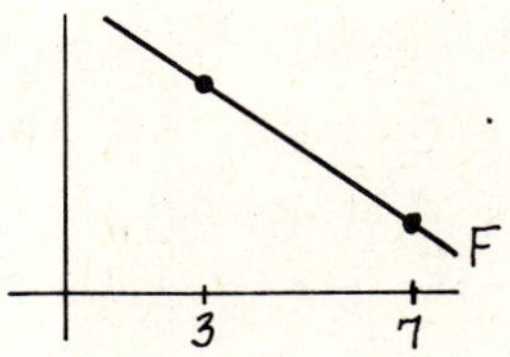

We found the coordinates of these points in example 3.

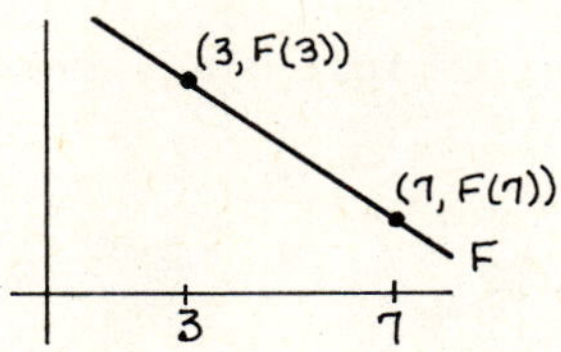

$$\text{Subtract}\ \begin{array}{r}(7,F(7))\\ \underline{(3,F(3))}\\ 4,F(7) - F(3)\end{array}.$$

$$\text{Slope} = \frac{\Delta y}{\Delta x} = \boxed{\frac{F(7) - F(3)}{4}}\ .$$

We can't simplify this unless we have a formula for F, as in the last example.

EXAMPLE 9 Find the slope of the following line by finding Δy and Δx between the indicated points.

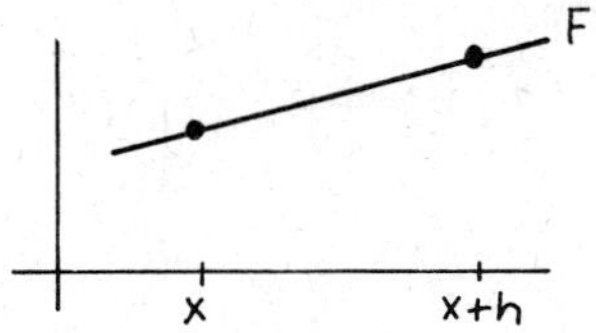

We found the coordinates of these points in example 4.

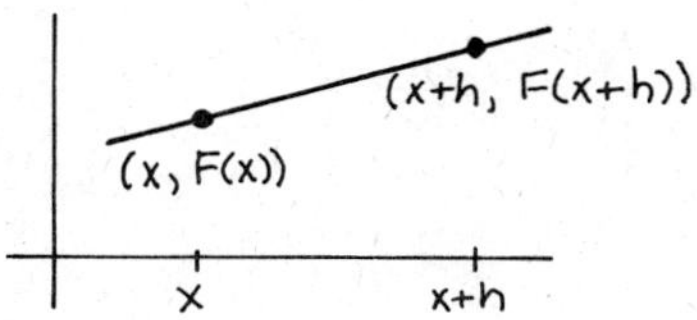

$$\begin{array}{rl} & (x + h, F(x + h)) \\ \text{Subtract} & \underline{(x \quad\;\;, F(x) \qquad\;)} \\ & h, F(x + h) - F(x) \end{array}.$$

$$\text{Slope} = \frac{\Delta y}{\Delta x} = \boxed{\frac{F(x + h) - F(x)}{h}}.$$

We can't simplify this unless we know a formula for F. The next example illustrates this.

EXAMPLE 10 Find the slope of the following lines by finding Δy and Δx between the indicated points.

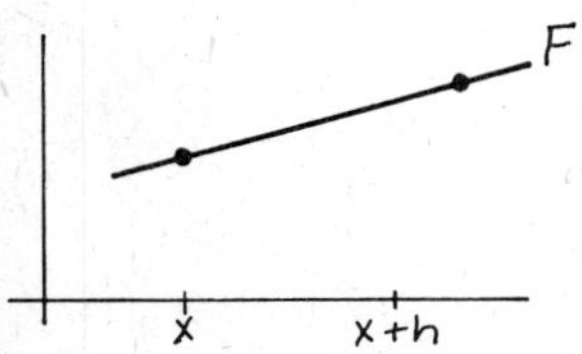

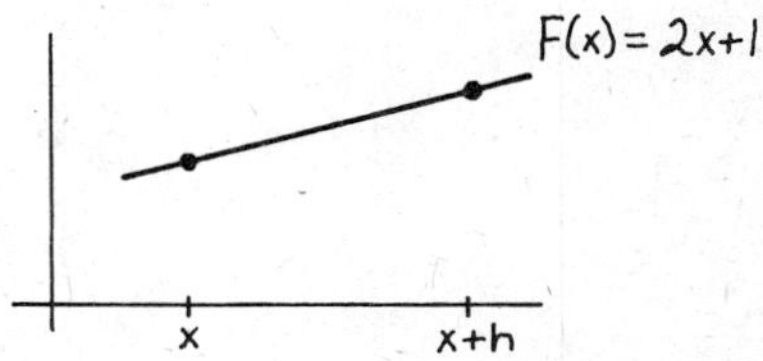

These two problems are the same except that in the right-hand one we have a formula for the function, so we can calculate the outputs:

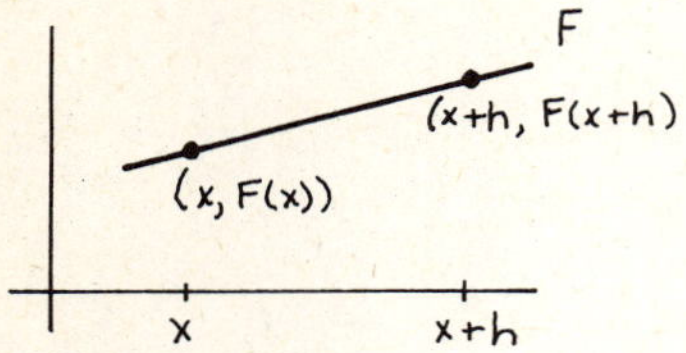

(x + h, F(x + h))

Subtract (x , F(x)).

h, F(x + h) - F(x)

$$\text{Slope} = \frac{\Delta y}{\Delta x} = \boxed{\frac{F(x + h) - F(x)}{h}}.$$

F(x) = 2x + 1

F( ) = 2( ) + 1

F(x + h) = 2(x + h) + 1 = 2x + 2h + 1

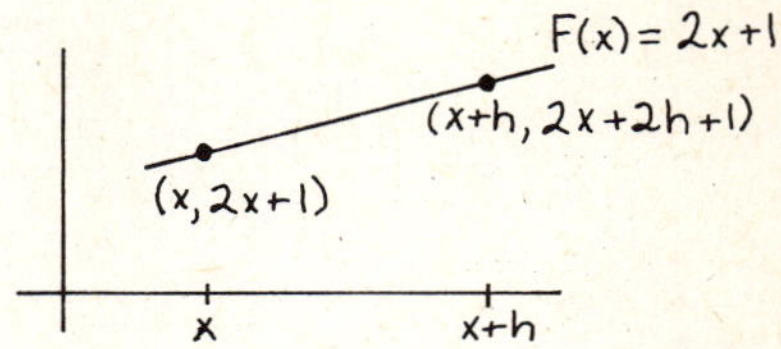

(x + h, 2x + 2h + 1)

Subtract (x , 2x + 1).

h, 2h

$$\text{Slope} = \frac{\Delta y}{\Delta x} = \frac{2h}{h} = \boxed{2}.$$

Notice the similarity between the steps of both solutions.

<u>EXAMPLE 11</u> Find the slope of the straight line that connects the indicated points on the curve.

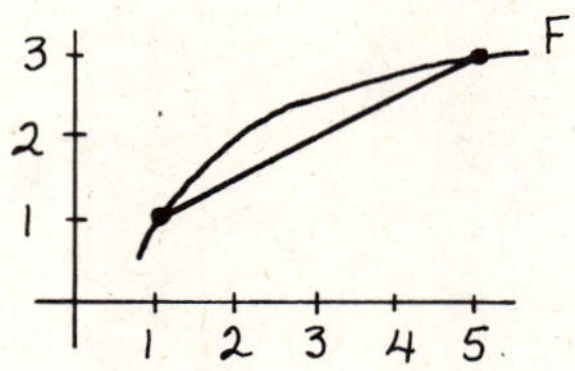

In this case, the curve is only a distraction. We can find the coordinates of the points directly from the numbers along the axes, as in examples 1 and 6.

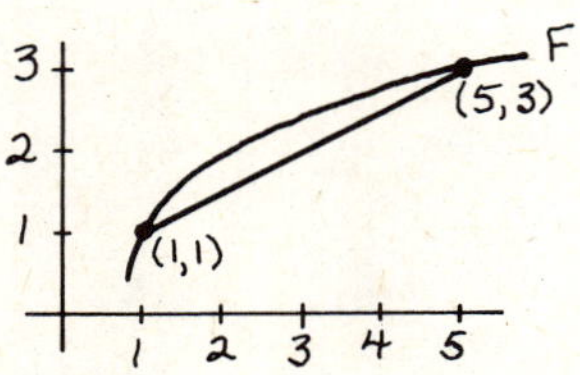

(5,3)

Subtract (1,1).

4,2

$$\text{Slope} = \frac{\Delta y}{\Delta x} = \frac{2}{4} = \boxed{\frac{1}{2}}.$$

EXAMPLE 12 Find the slope of the straight line that connects the indicated points on the curve.

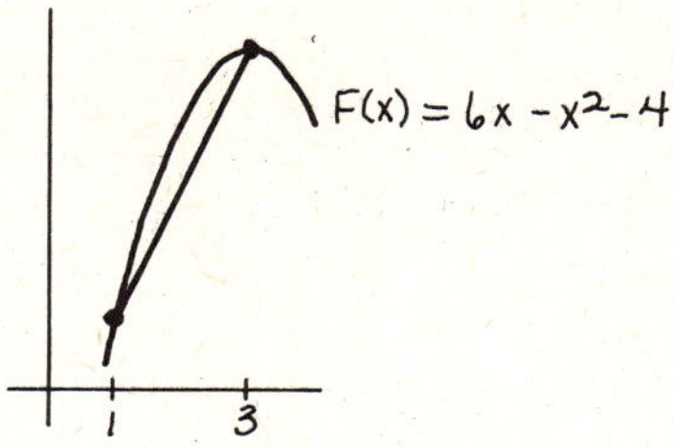

We can use the formula of F to compute the outputs.

F(1) = 1, so the coordinates of the left-hand point are (1,1).

F(3) = 5, so the coordinates of the right-hand point are (3,5).

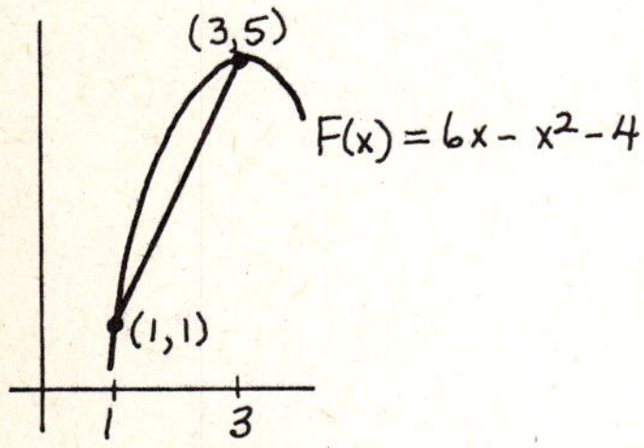

Subtract
$$\begin{array}{r} (3,5) \\ \underline{(1,1)} \\ 2,4 \end{array}$$
. Slope $= \frac{\Delta y}{\Delta x} = \frac{4}{2} = \boxed{2}$.

EXAMPLE 13 Find the slope of the straight line that connects the indicated points on the curve.

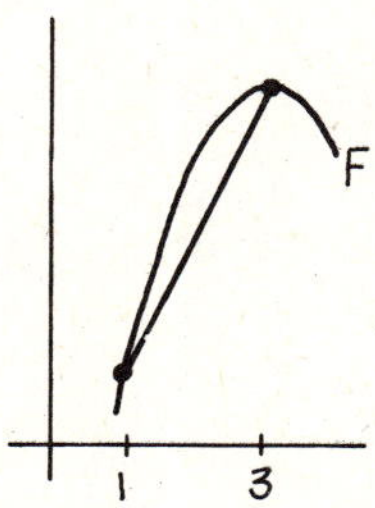

This time there is no formula for F, so the best we can do for the coordinates of the indicated points is:

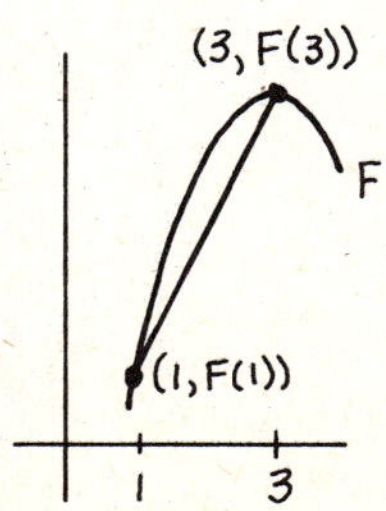

Subtract
$$\begin{array}{r} (3,F(3)) \\ \underline{(1,F(1))} \\ 2,F(3) - F(1) \end{array}$$
.

Slope $= \frac{\Delta y}{\Delta x} = \boxed{\frac{F(3) - F(1)}{2}}$.

EXAMPLE 14 Find the slope of the straight line that connects the indicated points on the curve.

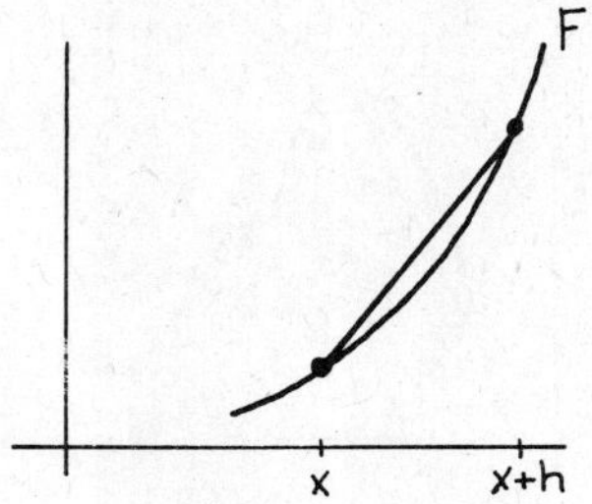

This is like example 9. The best answer we can give for the coordinates of the indicated points is:

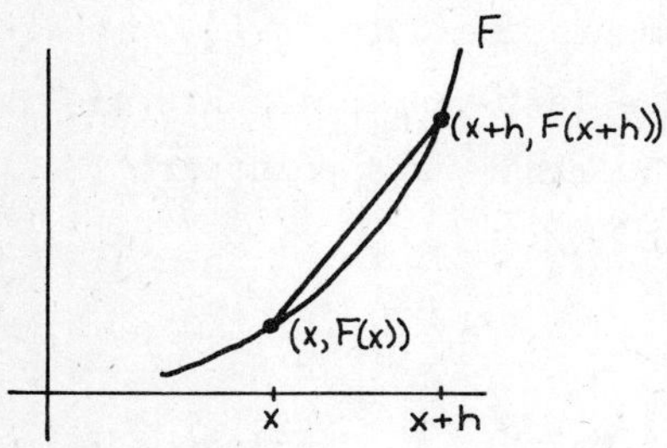

$$\begin{array}{rl} & (x + h, F(x + h)) \\ \text{Subtract} & \underline{(x\qquad, F(x)\qquad)} \, . \\ & \qquad h, F(x + h) - F(x) \end{array}$$

$$\text{Slope} = \frac{\Delta y}{\Delta x} = \boxed{\frac{F(x + h) - F(x)}{h}} \, .$$

This is the best answer we can give, unless we know a formula for F. Notice that this answer is exactly the same as the answer to example 9, even though the graphs are entirely different. No matter what the graph of F looks like, the slope of the line connecting two points on the graph is $\frac{F(x + h) - F(x)}{h}$. If we know a formula for F, we can simplify this.

EXAMPLE 15 In each case, find the slope of the straight line that connects the indicated points on the curve.

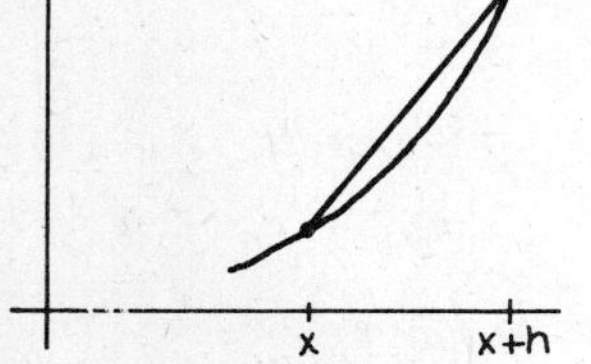

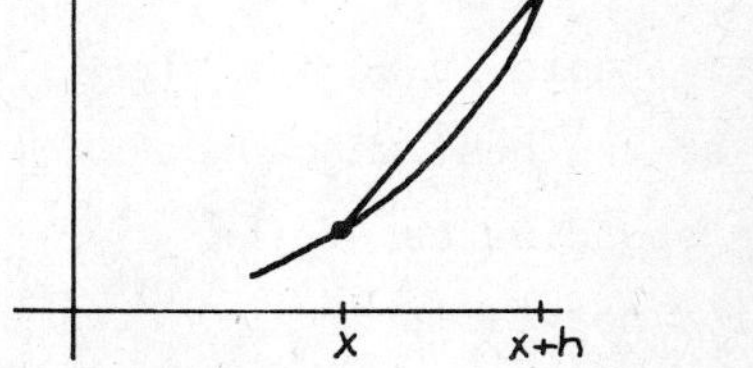

These two problems are the same, except that in the right-hand one we have a formula for the function, so we can calculate the outputs:

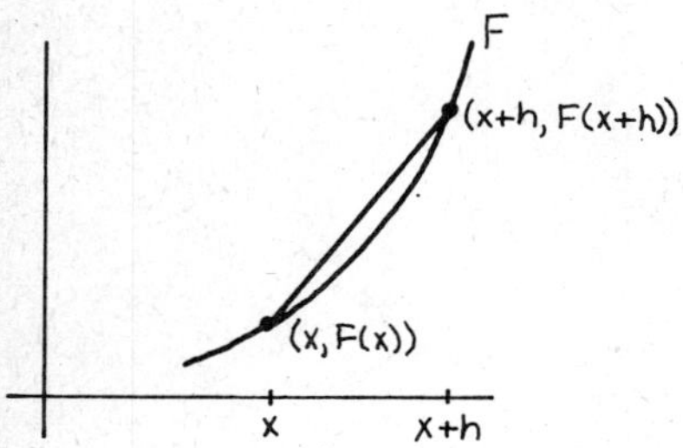

$$\begin{array}{rl} & (x+h, F(x+h)) \\ \text{Subtract} & \underline{(x \quad\;\; , F(x) \quad\;\;)} \\ & \quad\;\; h, F(x+h) - F(x) \end{array}.$$

$$\text{Slope} = \frac{\Delta y}{\Delta x} = \boxed{\frac{F(x+h) - F(x)}{h}}.$$

$$F(x) = x^2$$

$$F(\qquad) = (\qquad)^2$$

$$F(x+h) = (x+h)^2 = x^2 + 2xh + h^2$$

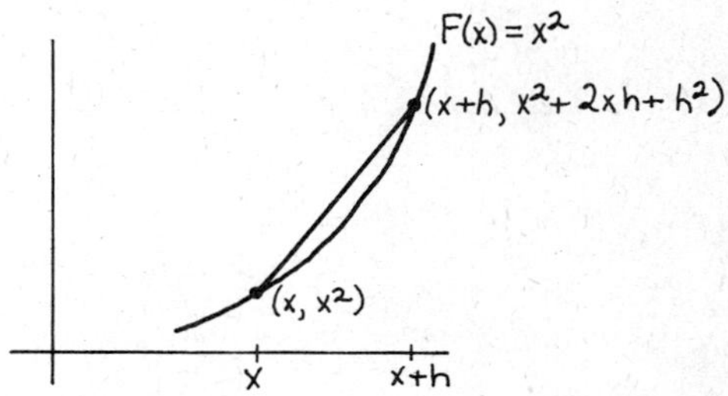

$$\begin{array}{rl} & (x+h, x^2 + 2xh + h^2) \\ \text{Subtract} & \underline{(x \quad\;\; , x^2 \qquad\qquad\quad)} \\ & \quad\;\; h, \qquad 2xh + h^2 \end{array}.$$

$$\text{Slope} = \frac{\Delta y}{\Delta x} = \frac{2xh + h^2}{h}$$

$$= \frac{h(2x+h)}{h} = \boxed{2x + h}.$$

Notice the similarity between the steps of both solutions.

The same problem is sometimes presented this way:

<u>EXAMPLE 16</u> In each case, find the slope of the straight line that connects the indicated points on the curve.

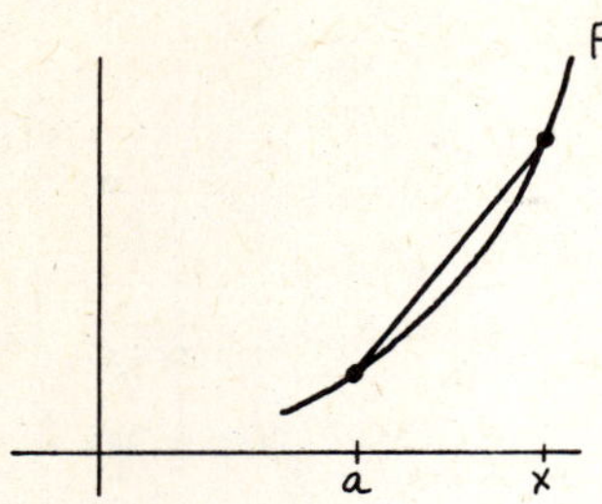

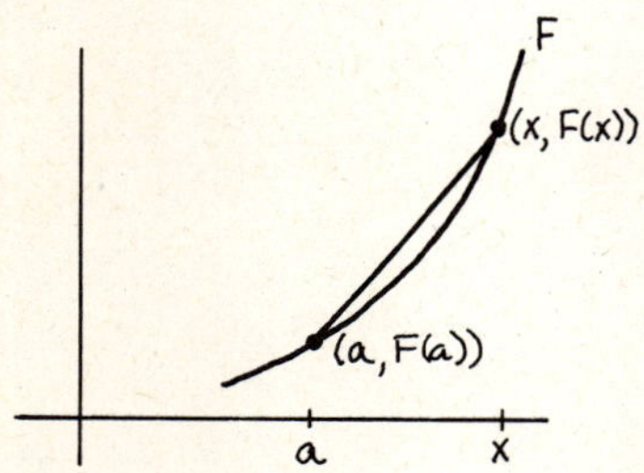

$$\begin{array}{rl} & (x, F(x)) \\ \text{Subtract} & \underline{(a, F(a))} \\ & x - a, F(x) - F(a) \end{array}$$

$$\text{Slope} = \boxed{\frac{F(x) - F(a)}{x - a}}.$$

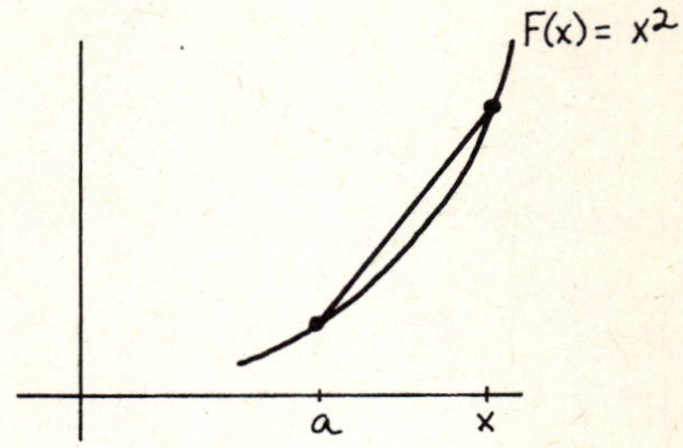

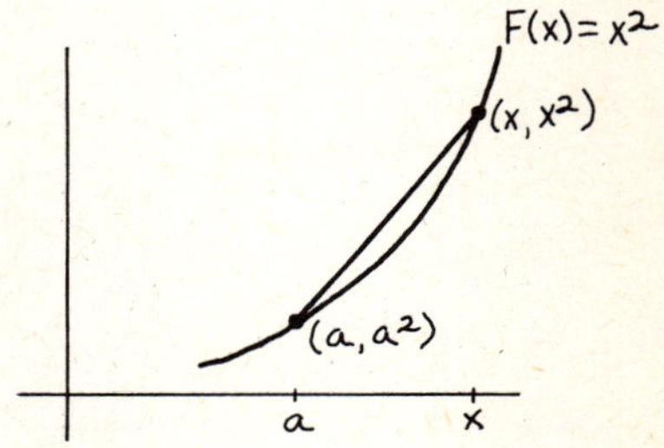

$$\begin{array}{rl} & (x, x^2) \\ \text{Subtract} & \underline{(a, a^2)} \\ & x - a, x^2 - a^2 \end{array}$$

$$\text{Slope} = \frac{x^2 - a^2}{x - a}$$

$$= \frac{(x + a)(x - a)}{(x - a)} = x + a.$$

Again, notice the similarity between the steps of both solutions. This example was the same as example 15, except that we used different letters to stand for the inputs.

Here is one more example of this type.

<u>EXAMPLE 17</u> Find the slope of the straight line that connects the indicated points on the curve.

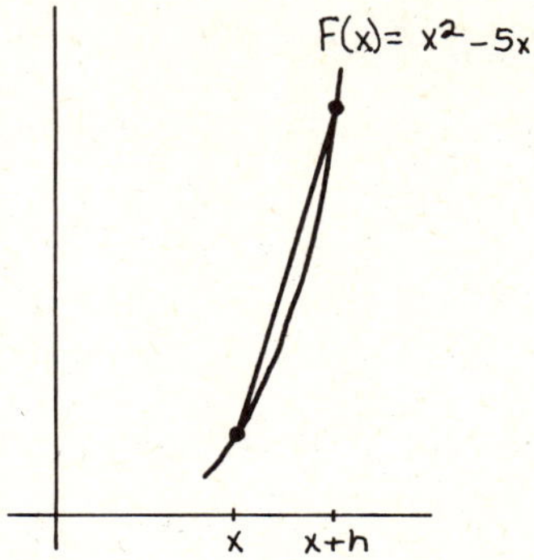

$$F(x) = x^2 - 5x$$

$$F(\quad) = (\quad)^2 - 5(\quad)$$

$$F(x+h) = (x+h)^2 - 5(x+h)$$

$$= x^2 + 2xh + h^2 - 5x - 5h$$

So the coordinates are:

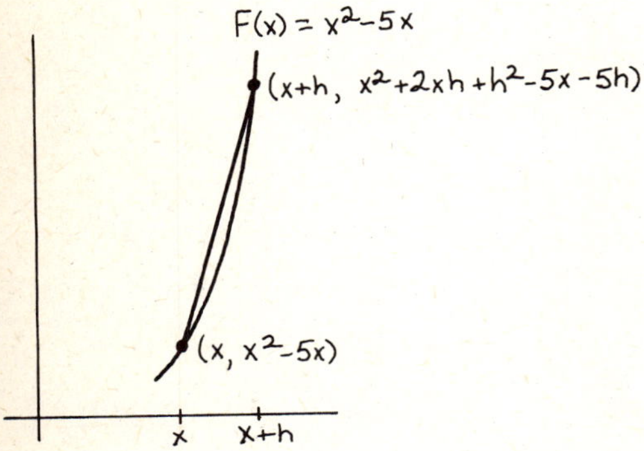

$$\begin{array}{lllll} & (x+h, & x^2+2xh+h^2 & -5x & -5h) \\ \text{Subtract:} & \underline{(x\quad ,} & \underline{x^2} & \underline{-5x} & \underline{\quad)}. \\ & h, & 2xh+h^2 & & -5h \end{array}$$

$$\text{Slope} = \frac{\Delta y}{\Delta x} = \frac{2xh+h^2-5h}{h} = \frac{h(2x+h-5)}{h}$$

$$= \boxed{2x+h-5}.$$

AREA

EXAMPLE 18 What is the area of each rectangle?

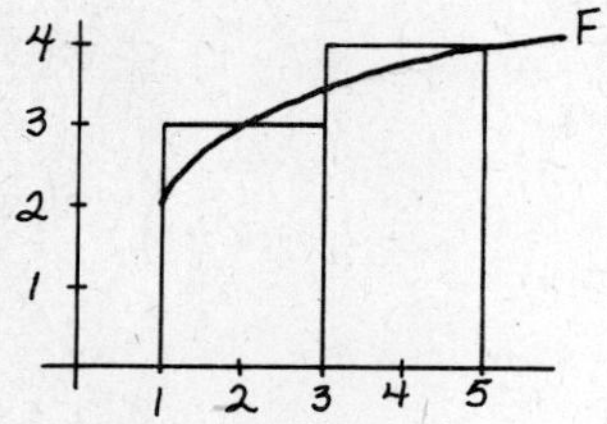

The left-hand rectangle has a height of 3 and a width of 2.

Area = height•width = 3•2 = 6.

The right-hand rectangle has a height of 4 and a width of 2.

Area = height•width = 4•2 = 8.

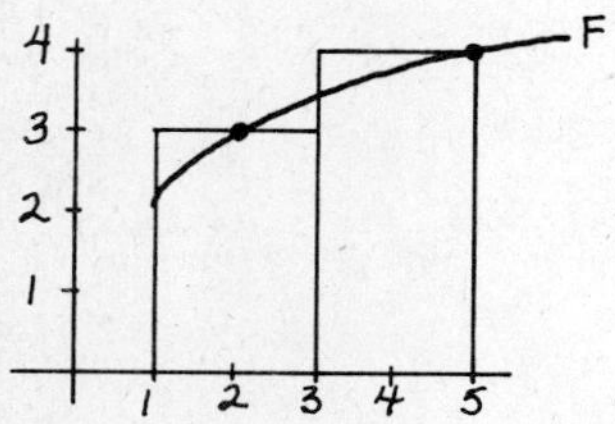

Notice that the height of the left-hand rectangle is the output of F given input 2, and the height of the right-hand rectangle is the output of F given input 5.

You can think of using the F(x)-axis as a ruler to measure these heights.

EXAMPLE 19 What is the area of each rectangle?

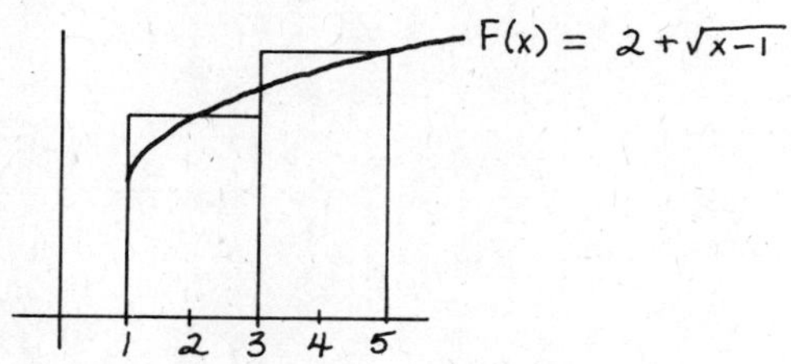

This time there are no numbers along the F(x)-axis, so we will have to use the formula for F to figure the heights.

Just as in the last example, the height of each rectangle is the output of F at the point where the top of the rectangle meets the graph.

The <u>height</u> of the left-hand rectangle is the output of F at this point.

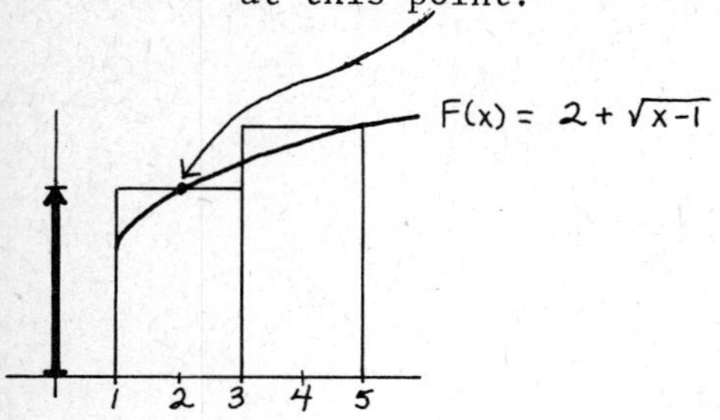

At the indicated point, the input is 2. F(2) = 3, so the height of the left-hand rectangle is 3.
Area = height•width = 3•2 = 6.

The <u>height</u> of the right-hand rectangle is the output of F at this point.

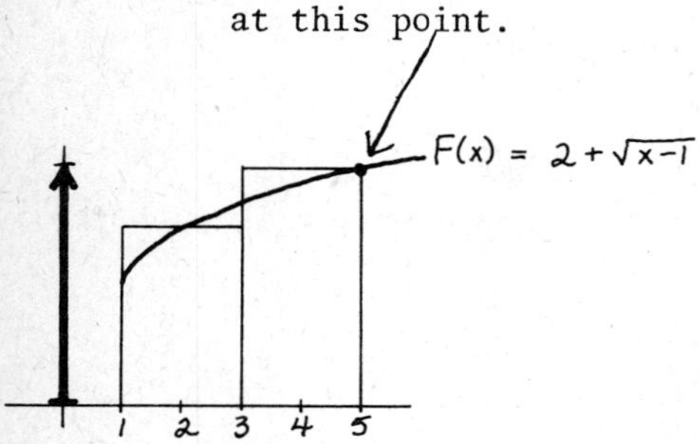

At the indicated point, the input is 5. But F(5) = 4, so the height of the right-hand rectangle is 4.
Area = height•width = 4•2 = 8.

<u>EXAMPLE 20</u> What is the area of each rectangle?

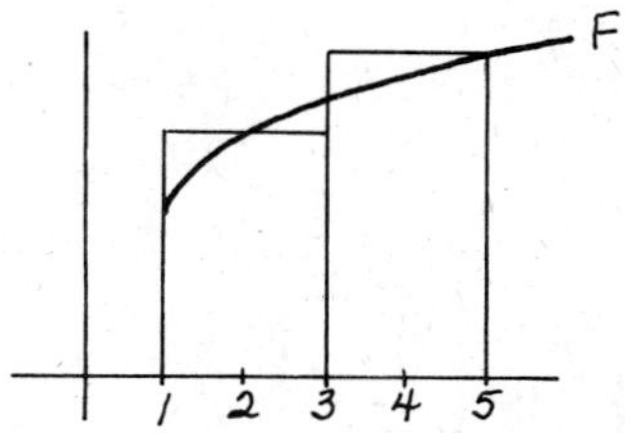

The height of the left-hand rectangle is the output of F given input 2. But with no numbers along the F(x)-axis and no formula for F, the best answer we can give for this output is F(2). So the height of the left-hand rectangle is F(2), and its width is 3 - 1 = 2.
Area = height•width = F(2)•2.

The height of the right-hand rectangle is F(5). The width is 5 - 3 = 2.
Area = height•width = F(5)•2.

These are the best answers we can give without knowing a formula for F. Notice that we figured the width by taking the input at the right side of the rectangle, and subtracting the input at the left side. This is like finding the horizontal separation, or Δx, between two points.

EXAMPLE 21 What is the area of each rectangle?

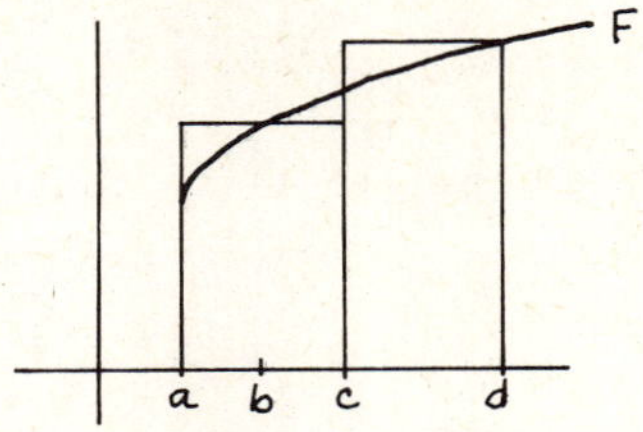

This graph has letters that stand for inputs. Otherwise, it is just the same as the last example.

The height of the left-hand rectangle is F(b). Its width is c - a.
Area = height•width = F(b)•(c - a).

The height of the right-hand rectangle is F(d). Its width is d - c.
Area = height•width = F(d)•(d - c).

EXAMPLE 22 What is the area of each rectangle?

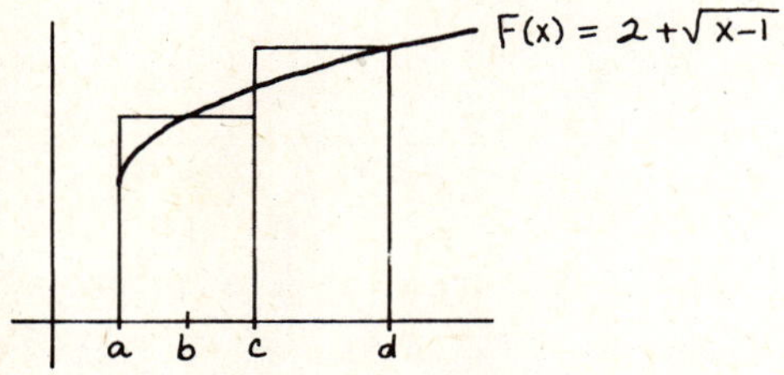

The height of the left-hand rectangle is F(b), but now we can use the formula for F. $F(b) = 2 + \sqrt{b - 1}$. The width is c - a.
Area = height•width = $(2 + \sqrt{b - 1})(c - a)$.

The height of the right-hand rectangle is F(d), which is $2 + \sqrt{d - 1}$.
The width is d - c. Area = height width = $(2 + \sqrt{d - 1})(d - c)$.

Note: In the answers to the last two examples, we needed all the parentheses. F(b)•(c - a) is correct, but F(b)•c - a means something different, and would not be a correct answer to example 19. $2 + \sqrt{b - 1}(c - a)$ would be wrong as an answer to example 20. In each case, we have to multiply the whole height times the whole width.

Finally, here is one more example of a question that math books may ask about abstract graphs.

MAXIMUMS AND MINIMUMS

EXAMPLE 23 For each function graphed below, find the maximum and minimum outputs and the corresponding inputs.

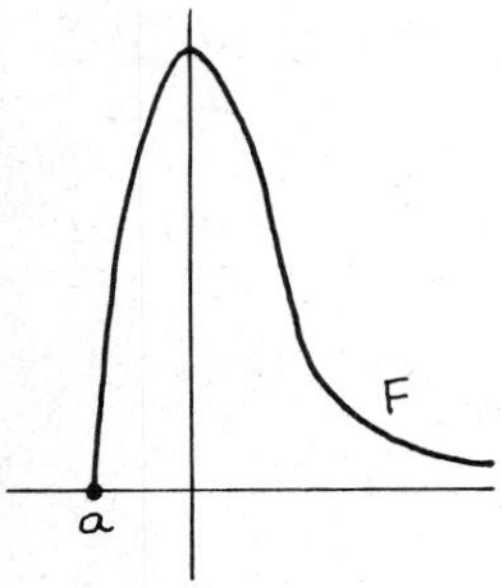

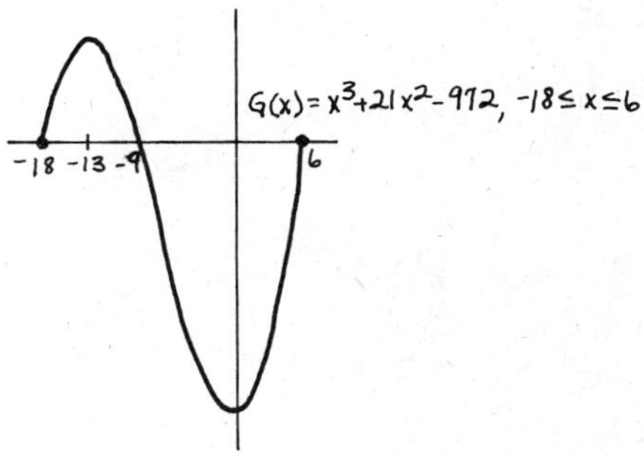

From the graph of F, we can see that the greatest output occurs when the input is 0. But we don't have a formula for F, so the best answer we can give for this output is F(0). The least output occurs when the input is a. Since this point is right on the x-axis, we know that the output there is 0.

rule F	greatest output:	F(0)	least output:	0
	corresponding input:	0	corresponding input:	a

From the graph of G, we can see that the greatest output occurs when the input is -13. We can figure the output, since there is a formula for G. $G(-13) = (-13)^3 + 21(-13)^2 - 972 = 380$. The least output occurs when the input is 0. G(0) = -972.

rule G	greatest output:	380	least output:	-972
	corresponding input:	-13	corresponding input:	0

EXERCISES (ANSWERS BEGIN ON PAGE 160.)

In problems 1-4, indicate the highest and lowest points on the graph, and give the coordinates of these points.

1.

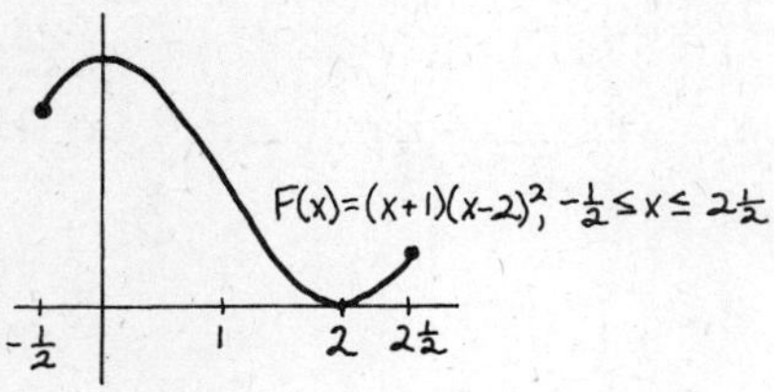

2.

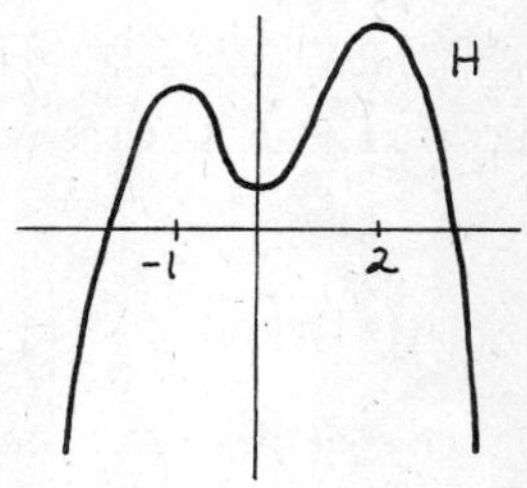

3.

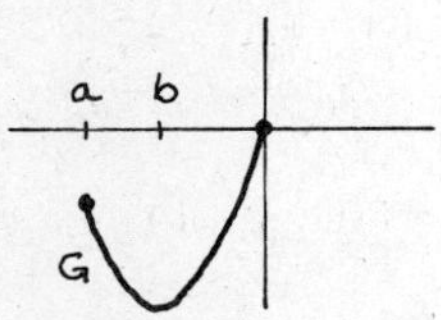

4.

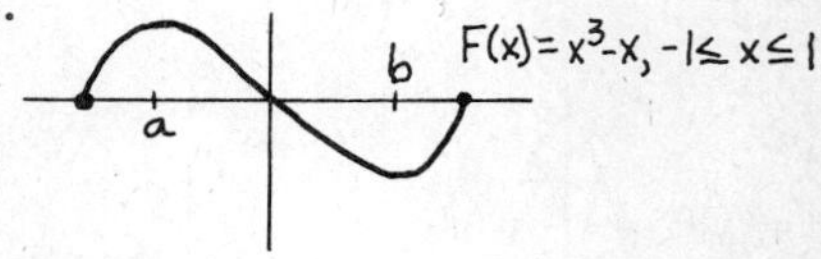

In problems 5-9, find the slope of the straight line that connects the two points on the curve.

5.

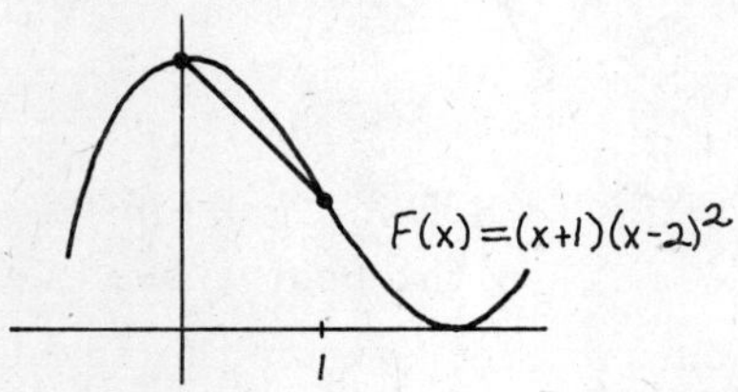

6.

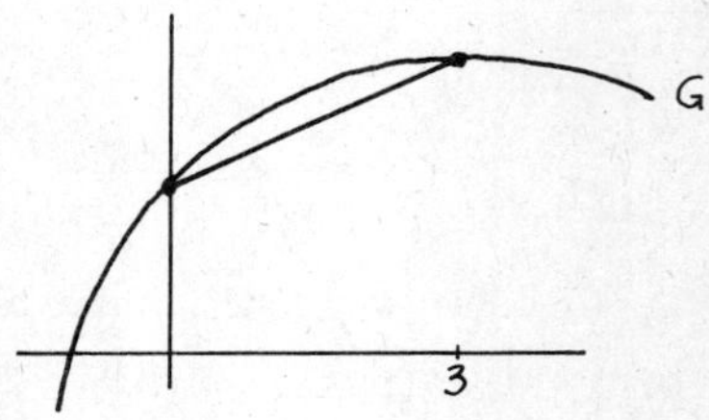

7.

8.

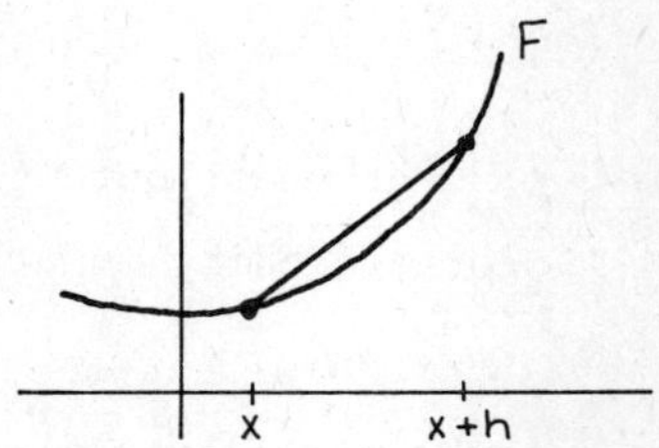

9.

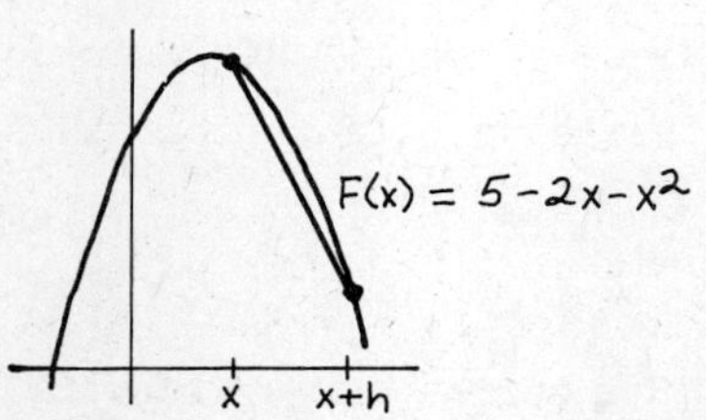

In problems 10-13, find the area of each rectangle.

10.

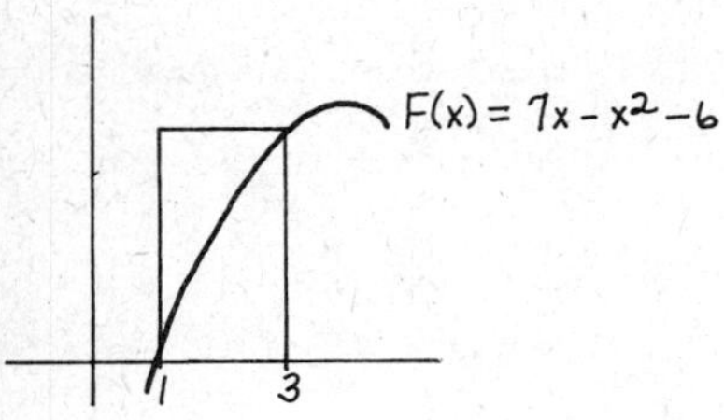

11.

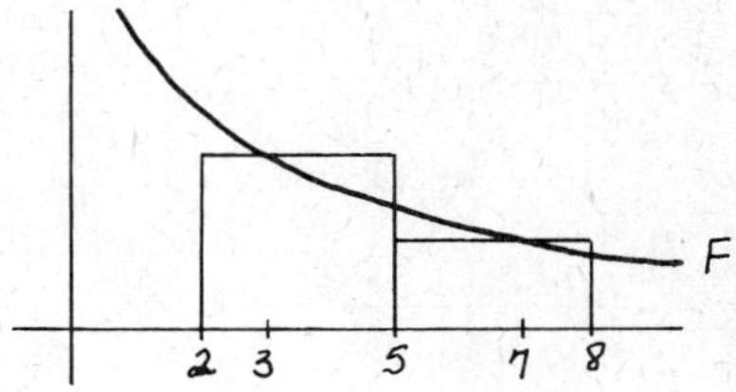

12.

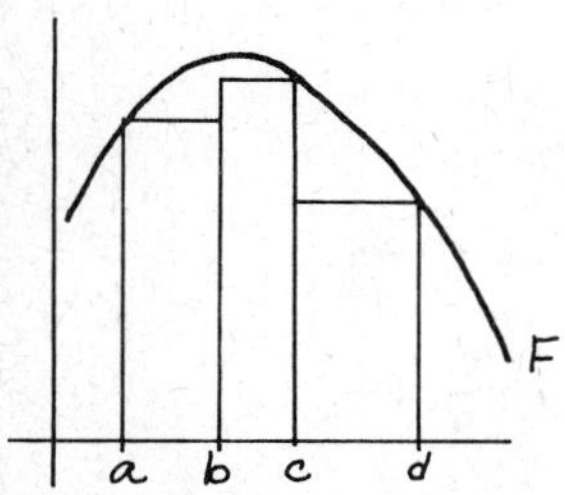

13.

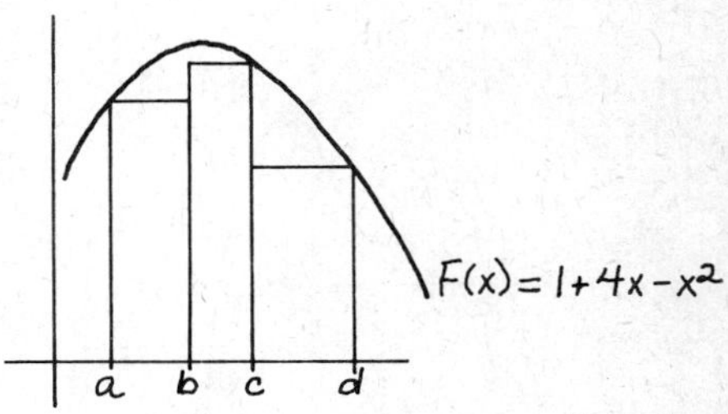

ANSWERS TO EXERCISES

(1)

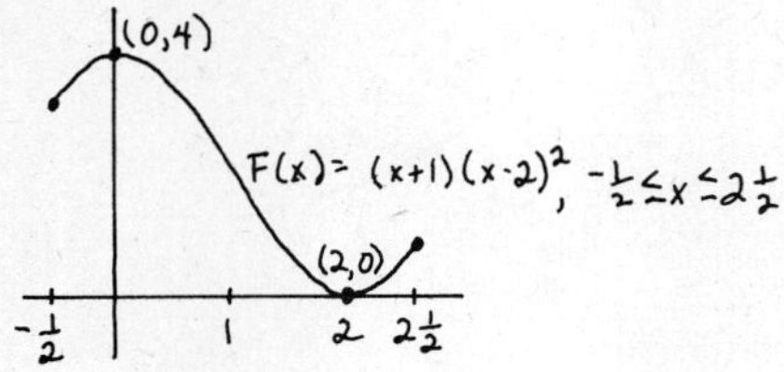

At the highest point, the input is 0.
The output is $F(0) = (1)(-2)^2 = 4$.

At the lowest point, the input is 2.
The output is $F(2) = (3)(0)^2 = 0$.

(2)

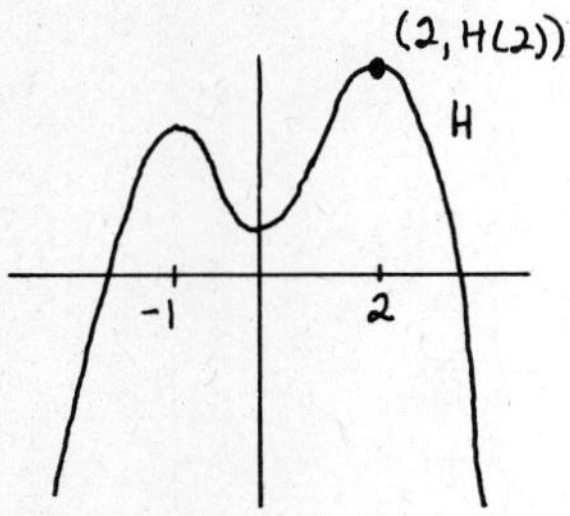

At the highest point, the input is 2.
The output is $H(2)$.

There is no lowest point.

(3)

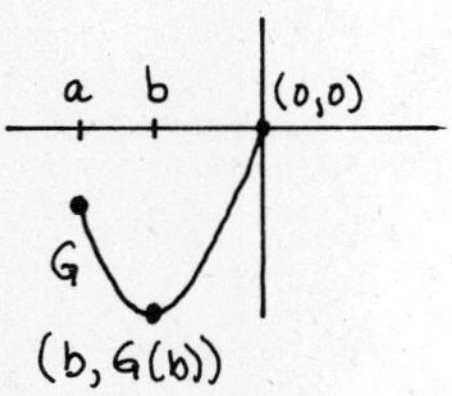

At the highest point, the input is 0. As we can see from the graph, the output is also 0.
At the lowest point, the input is b.
The output is $G(b)$.

(4)

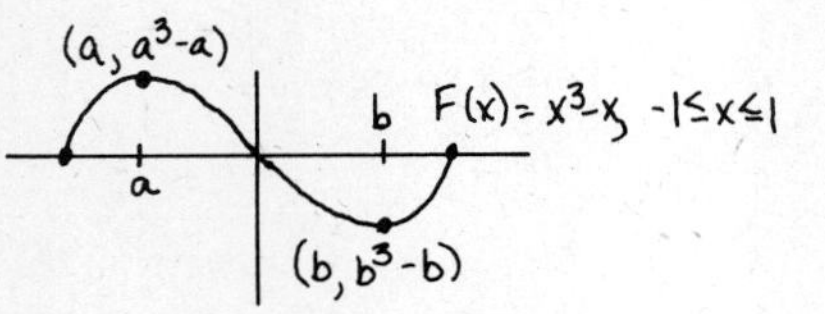

At the highest point, the input is a.
The output is $a^3 - a$.

At the lowest point, the input is b.
The output is $b^3 - b$.

(5)

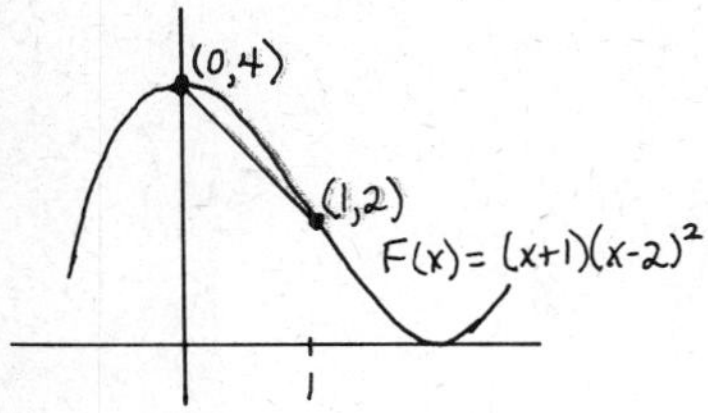

(1, 2)
(0, 4)
1, -2

slope $= \frac{-2}{1} = -2$

(6)

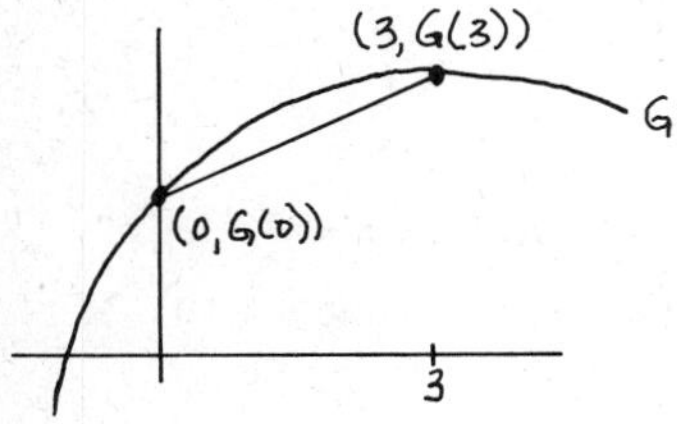

(3, G(3))
(0, G(0))
3, G(3) - G(0)

slope $= \frac{G(3) - G(0)}{3}$

(7)

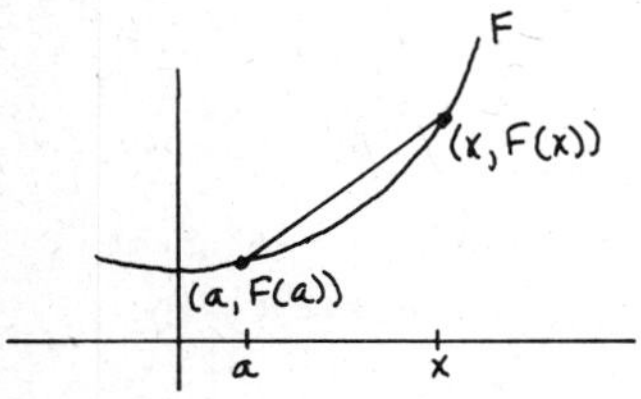

(x, F(x))
(a, F(a))
x - a, F(x) - F(a)

slope $= \frac{F(x) - F(a)}{x - a}$

(8)

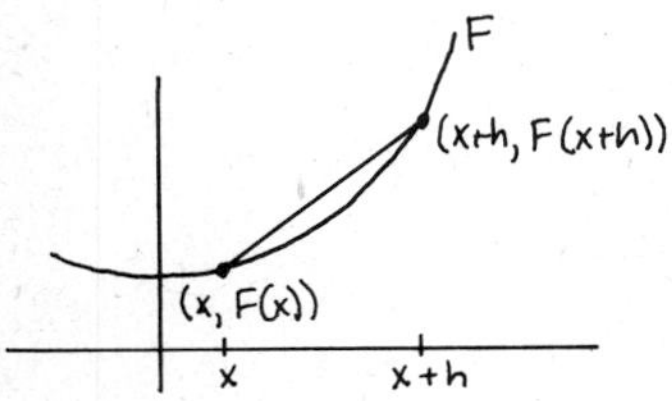

(x + h, F(x + h))
(x, F(x))
h, F(x+h) - F(x)

slope $= \frac{F(x+h) - F(x)}{h}$

(9)

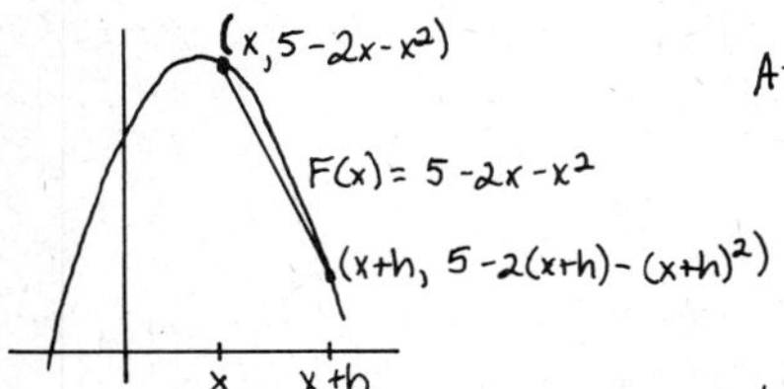

At the right hand point, the input is x+h.

$F(\quad) = 5 - 2(\quad) - (\quad)^2$

$F(x+h) = 5 - 2(x+h) - (x+h)^2$

$= 5 - 2x - 2h - (x^2 + 2xh + h^2)$

$= 5 - 2x - 2h - x^2 - 2xh - h^2$

(x + h, 5 - 2x - 2h - x² - 2xh - h²)
(x, 5 - 2x - x²)
h, -2h -2xh - h²

slope $= \frac{-2h - 2xh - h^2}{h} = \frac{h(-2 - 2x - h)}{h} = -2 - 2x - h$

(10)

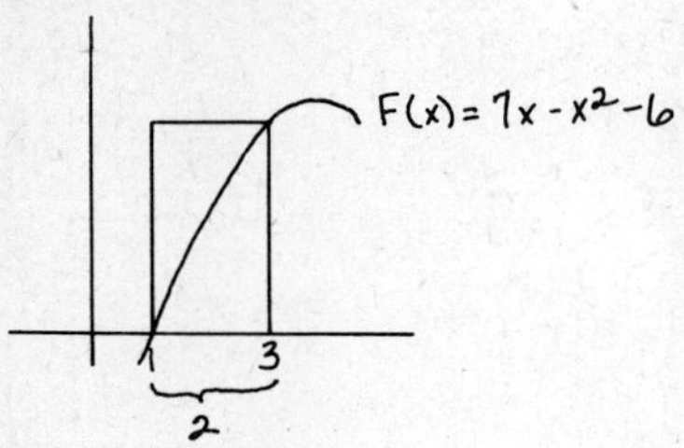

height $= F(3) = 7 \cdot 3 - 3^2 - 6 = 6$

width $= 3 - 1 = 2$

area $= 6 \cdot 2 = 12$

(11)

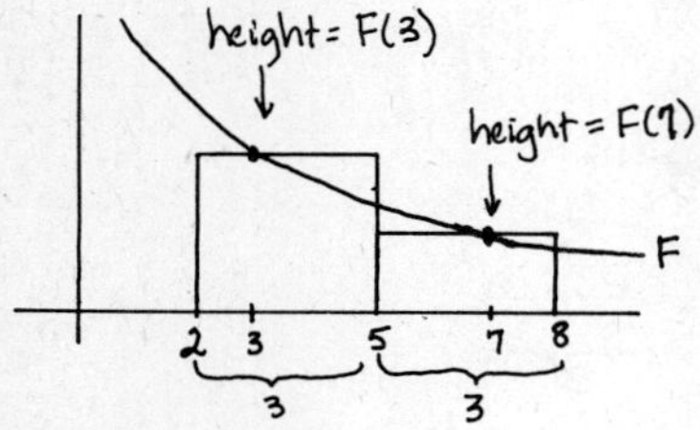

area of left hand rectangle $= F(3) \cdot 3$

area of right hand rectangle $= F(7) \cdot 3$

(12)

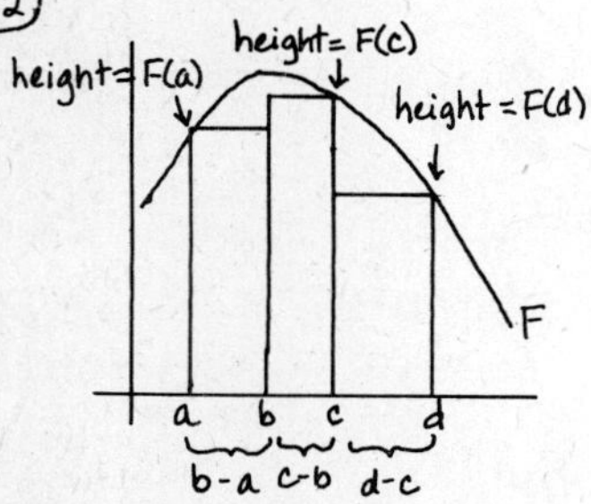

area of left hand rectangle $= F(a)(b-a)$

area of middle rectangle $= F(c)(c-b)$

area of right hand rectangle $= F(d)(d-c)$

(13)

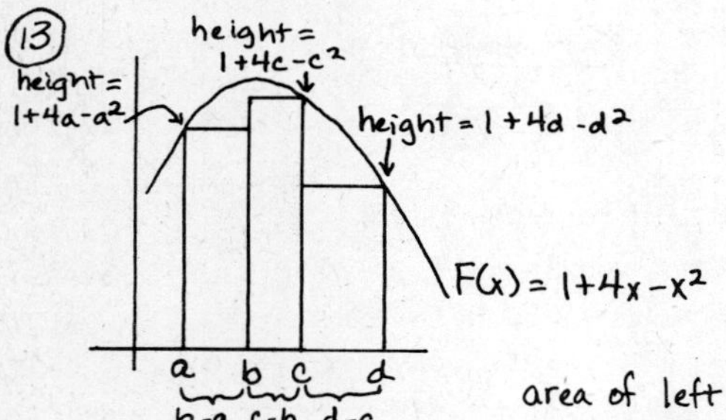

area of left hand rectangle $= (1+4a-a^2)(b-a)$

area of middle rectangle $= (1+4c-c^2)(c-b)$

area of right hand rectangle $= (1+4d-d^2)(d-c)$

ADDITIONAL EXERCISES

In problems 1-4, indicate the highest and lowest points on the graph, and give the coordinates of these points.

1.

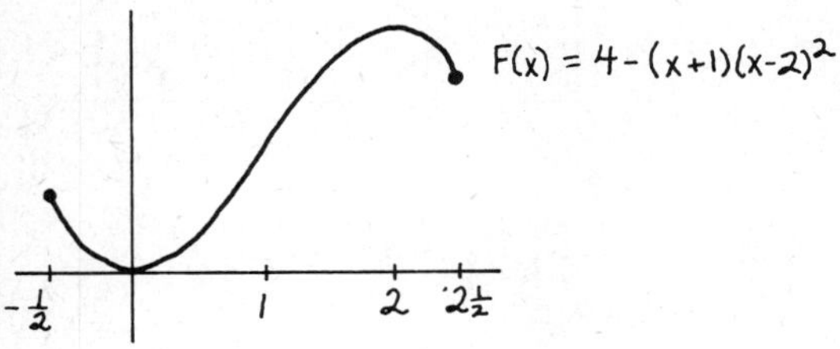

2.

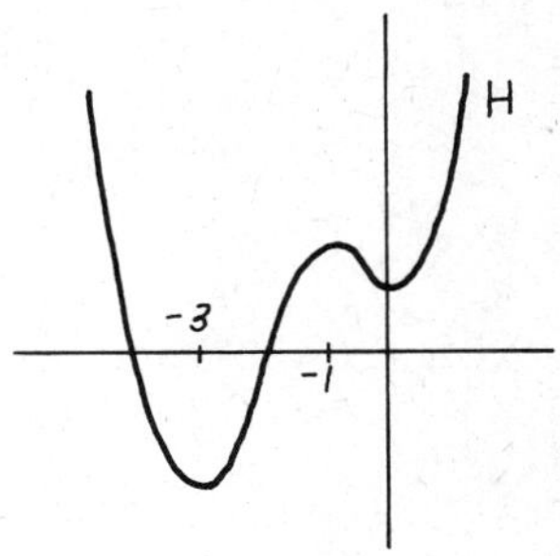

3.

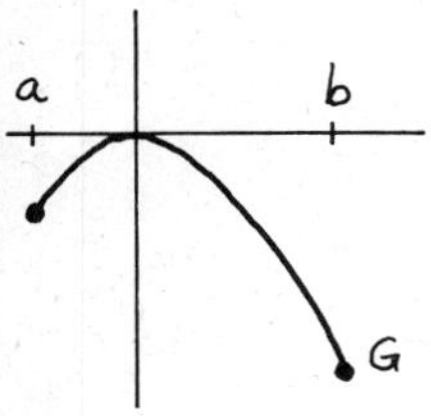

4.

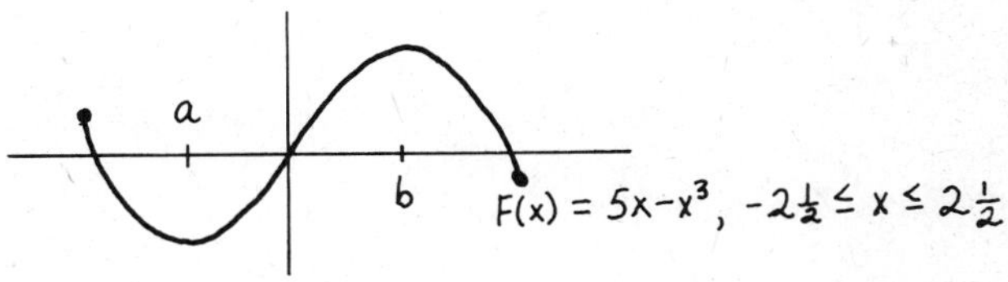

In problems 5-9, find the slope of the straight line that connects the two points on the curve.

5.

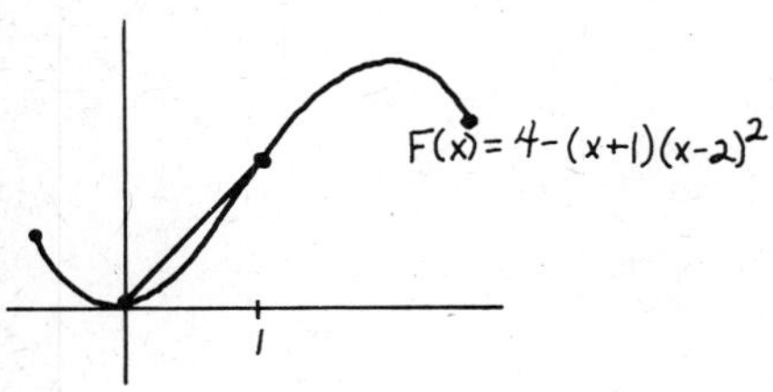

6.

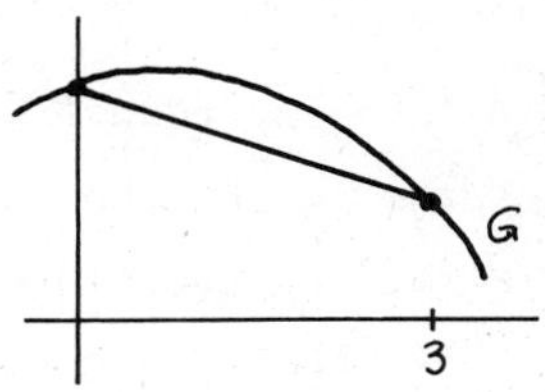

7.

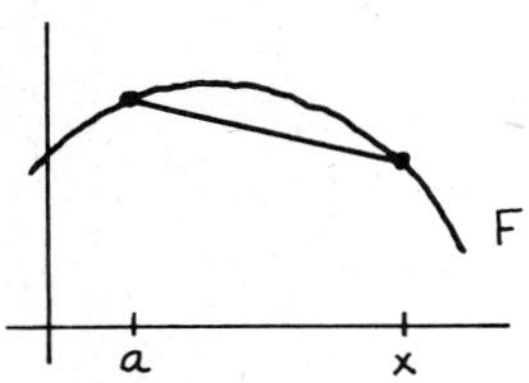

8.

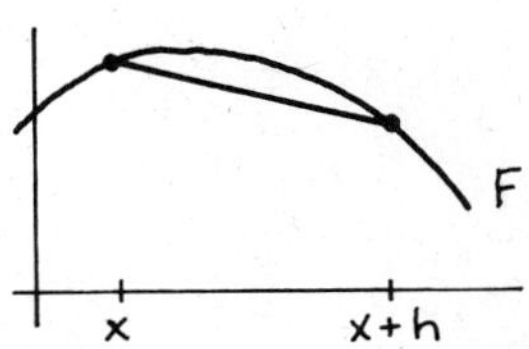

9.

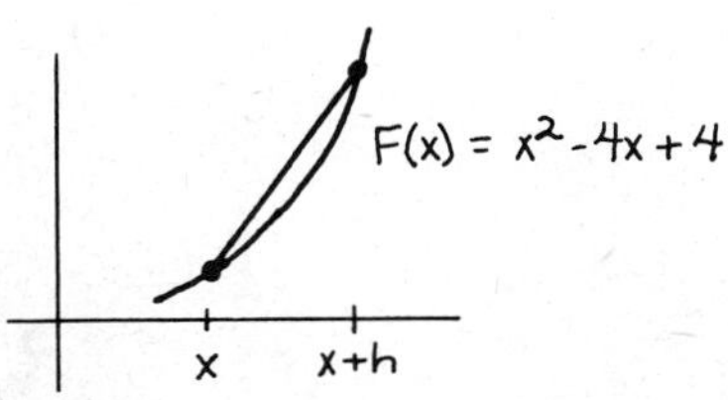

In problems 10-13, find the area of each rectangle.

10.

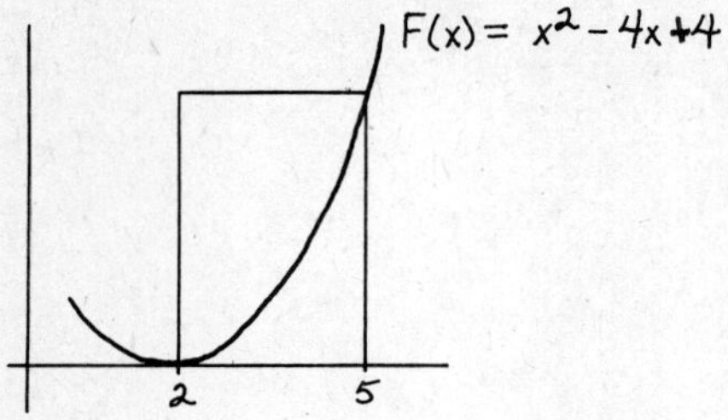

11.

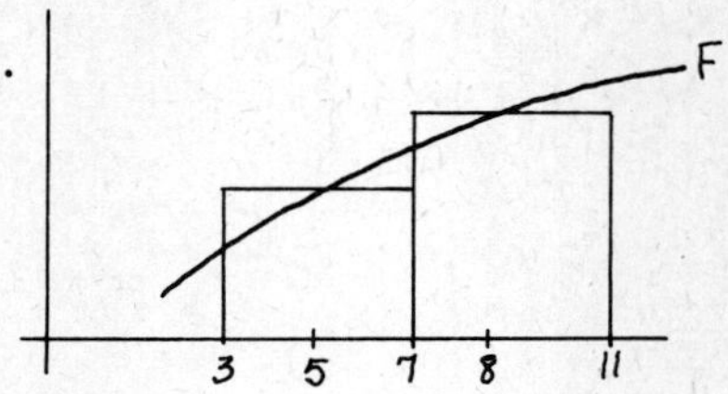

12.

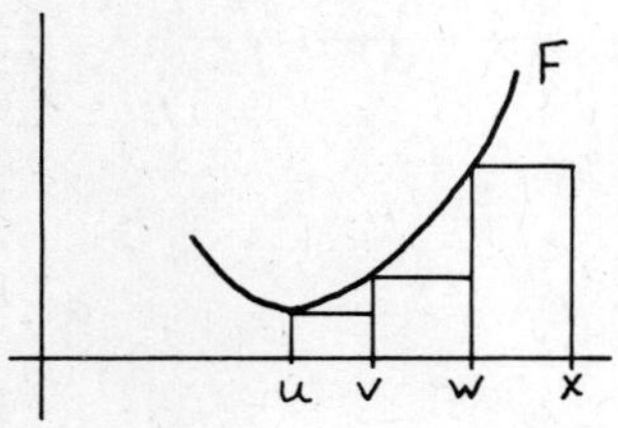

13.

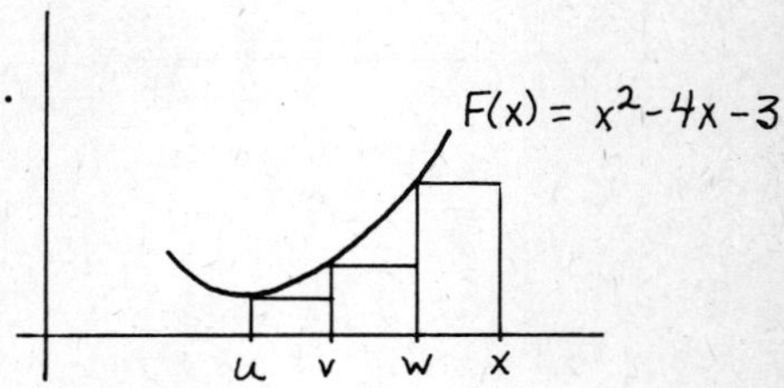

SUMMARY OF LESSON 8

Graphs in advanced math books are sometimes "abstract." They may use letters to stand for inputs, they often leave the F(x)-axis unnumbered, and they may present graphs of functions for which there are no formulas. Calculus courses usually require the following skills with such graphs:

- ✓ labeling points on the graph [pages 141-142], especially the highest and lowest points [page 157].
- ✓ finding the slope of a straight line [pages 144-148], especially of a line connecting two points on a curve [pages 148-153].
- ✓ finding the areas of rectangles under the graph of a function [pages 154-157].

If you have trouble with any of the exercises, and the answers don't give enough help, use the following as a guide for review.

exercises	review
1-4	example 23, page 157
1	example 2, page 142
2	example 3, page 142
3	example 4, page 143
4	example 5, page 143
5-9	example 6, page 144
5	example 12, page 149
6	example 13, page 149
7	example 16, (left-hand part), page 152
8	example 14, page 150
9	example 15, page 150; example 17, page 153
10	example 19, page 154
11	example 20, page 155
12	example 21, page 156
13	example 22, page 156

INDEX